T0137355

Lecture Notes in Social Networks

More information about this series at http://www.springer.com/series/8768

Yan Chen · H. Vicky Zhao

Behavior and Evolutionary Dynamics in Crowd Networks

An Evolutionary Game Approach

Yan Chen
School of Cyberspace Security
University of Science
and Technology of China
Hefei, China

H. Vicky Zhao
Department of Automation
Beijing National Research Center
for Information Science and Technology
Tsinghua University
Beijing, China

ISSN 2190-5428 ISSN 2190-5436 (electronic)
Lecture Notes in Social Networks
ISBN 978-981-15-7162-6 ISBN 978-981-15-7160-2 (eBook)
https://doi.org/10.1007/978-981-15-7160-2

This Springer imprint is published by the registered company Springer Nature Singapore Pte Ltd.
The registered company address is: 152 Beach Road, #21-01/04 Gateway East, Singapore 189721, Singapore

To our families.

Acknowledgements

We would like to thank The National Key Research and Development Program of China for their financial support (Project ID 2017YFB1400100). We would also like to thank Ms. Hangjing Zhang, Mr. Yuejiang Li, Mr. Hong Hu and Ms. Yaxin Li for their help during the preparation of this book.

Contents

List of Figures

List of Tables

Chapter 1
Introduction

Abstract In the emerging crowd cyber-eco systems, millions of deeply connected individuals, smart devices, government agencies, and enterprises actively interact with each other and influence each other's decisions. It is crucial to understand such intelligent entities' behaviors and to study their strategic interactions, which provides important guidelines on the design of reliable networks capable of predicting and preventing detrimental events with negative impacts on our society and economy. This chapter introduces basic concepts in behavior and evolutionary dynamics in crowd networks. Using information diffusion over social networks as an example, we discuss challenges in the modeling and analysis of user behavior and their interactions in large-scale, decentralized and heterogeneous networks, and introduce an evolutionary game theoretic framework to study behavior dynamics in crowd networks.

Keywords Crowd networks · Behavior dynamics · Information diffusion · Social networks · Evolutionary game theory

1.1 Crowd Networks

With the recent development of social media, Internet of things, big data, cloud computing, and many other new technologies, we witness new industrial and social management patterns, and the emergence of crowd cyber eco-systems consisting of smart and deeply connected entities such as individuals, enterprises and government agencies. Examples of such crowd networks include online e-commerce platforms such as Amazon and Tmall that connect customers, small businesses, and large enterprises, online communities such as Reddit and Quora that connect users worldwide with a common interest, globally connected supply chains, etc.

In these networks, smart entities regularly interact with each other, influence each other's decisions, and have a significant impact on our society and economy. Due to the complex structures of these networks and the dynamic interactions among users, we often observe dangerous and damaging events that spread rapidly and

Y. Chen and H. V. Zhao, *Behavior and Evolutionary Dynamics in Crowd Networks*, Lecture Notes in Social Networks, https://doi.org/10.1007/978-981-15-7160-2_1

globally in these networks [1]. One example is the "salt panic" in China after the 2011 Tohoku Tsunami, where the news of nuclear leakage greatly stimulated the rumors like "iodized salt can help ward off radiation poisoning", which lead to the "long lines and mob scenes at stores" and "10-fold jump of salt price" throughout China [2]. Another example is the COVID-19 crisis, where the highly contagious novel coronavirus spreads quickly all over the world and has severe impacts on our health, social lives, and economics [3]. Thus, to better understand the behavior of these complex crowd networks and to avoid such detrimental events, it is of crucial importance to study how such smart entities interact with each other, to understand their decision-making process, to analyze how they influence each other and the impact of such interactions on the entire crowd intelligence networks. Such investigation provides important guidelines on the design of efficient and effective mechanisms to manage such crowd intelligence networks.

1.2 Information Diffusion

In this book, we use information propagation over social networks as an example and study the impact of user behavior on network evolution. The social network, a social structure made up of social actors (such as individuals or organizations), sets of dyadic ties, and other social interactions between actors, is a prominent tool for the diffusion of information in society [4]. For example, by posting some videos and pictures related to personal preferences on Facebook, or declaring some political opinions on Twitter, or releasing promotional advertisements in Wechat Moments, people interact with each other for their own purposes. Once a piece of information is published on the network, it may vanish quickly after its appearance, or it may last for a long time and inspire a heated discussion. Thus, the prediction of its propagation process and the final destiny of information spread is vital in many applications, including advertising and political campaigns. Information diffusion over social networks may also have significant impacts on our society. As an example, according to MIT Technology Review, in the COVID-19 crisis, "social media has zipped information and disinformation around the world at unprecedented speeds, fueling panic, racism, ... and hope" [5]. Therefore, modeling and predicting information diffusion over social networks has been a hot research topic in recent years.

The research on information diffusion is critical. On the one hand, from the perspective of profits, the study of information diffusion can help the enterprises or politicians to identify influential users and links and then make advertisements or advocation more effective. On the other hand, from a security perspective, this study can help prevent the spread of detrimental information such as rumors and computer viruses, which would reduce unnecessary losses significantly. The recent developments of information and communication technologies enable us to collect, store, and access big data, making the research of information dissemination versatile and meaningful, but also more challenging.

The study on information diffusion originates from the research of computer virus/epidemic spreading over networks [6]. One of the earliest and prominent works about information diffusion is [7], which focused on the dynamics of information propagation through blog space from both macroscopic and microscopic points of views. Subsequently, there are a large number of works on information diffusion. The researchers explore the problem from different angles and adopt a variety of methods to resolve the problem. From the research object, existing works can be divided into three categories: (1) diffusion characteristics analysis; (2) diffusion dynamics analysis; (3) diffusion stability analysis. Based on the adopted method, works could also be divided into two categories. The first category models the information diffusion from the macroscopic aspect, usually adopting machine learning or data mining techniques to predict the dynamics or properties of networks. The methods based on machine learning or data mining have a common drawback–a lack of understanding of the underlying microscopic mechanisms of individual decision-making during the diffusion process, which is the focus of the second category. The second category stressed on micro exploration, paying more attention to the decisions and motivations of individuals.

In recent years, evolutionary game theory (EGT) has attracted lots of attention. It is used to fully understand the details of information diffusion process and simulate the entire process, including diffusion dynamics as well as the final result. Initially, EGT derives from a biological problem of how to explain ritualized animal behavior in a conflict situation and then has become of interest to economists, sociologists, anthropologists, and philosophers [8]. It addresses the shortcomings of traditional game theory. In classical game theory, players are required to make rational choices, which means they need to carefully consider complicated reasonings, such as what they want, what their opponents want, what their opponents know, etc., and determine the best strategy in the competitions. At the same time, in EGT, there is no limitation for players' actions and few assumptions about the reasoning processes of players, and the process of natural selection, i.e., evolution, is the focus.

1.3 Evolutionary Game Theory

EGT defines a framework of contests, strategies, and the measurable criteria that can be used to predict the performances of competing strategies. The results of a game include the dynamics of changes in the population, whether the strategy is successful, and any achievable equilibrium states. These basic elements in a game just correspond with the fundamental elements in information diffusion, so the whole information diffusion process could be regarded as a game: for users being players, their adopted strategies being the strategies in the game, the information spreading process being the evolution, and the consequence of information (survive or vanish, and if survive, how many users accept this piece of information) being the equilibrium states. Different from methods using lots of data, by applying EGT to information diffusion, we could predict every small change in the process, get the detailed dynamics, and finally foretell the stable states. Meanwhile, we can interpret the mechanisms of how users interact with others from the individual's view, rather than the whole network, which helps to understand the diffusion process more deeply.

In related research, the evolutionary game-theoretic models have been proven to achieve high accuracy, with less calculation than machine learning or data mining approaches. Based on the conclusions of the EGT framework, the dynamics and stable states in the process of information diffusion can be quickly predicted, which could be applied to plenty of areas such as online advertisements, rumor control, and network security. Therefore, in this book, we introduce several evolutionary game-theoretic models under different scenarios and illustrate how to apply EGT to the analysis of information diffusion. Evolutionary game theory has also been used to study user behavior in other applications, such as the natural biological world, the social world, and the virtual online world [9] as well as traffic flow analysis and epidemics [10].

1.4 About This Book

In this book, we aim to offer a holistic evolutionary game-theoretic framework to theoretically study behavior and evolutionary dynamics in large-scale, decentralized and heterogeneous crowd networks. We combine mathematical tools and engineering concepts with ideas from sociology, biology and game theory, and propose an interdisciplinary approach to model intelligent entities' decision-making process and their interactions. In this book, we review the fundamental methodologies to study user interactions and evolutionary dynamics in crowd networks and discuss recent advances in this emerging inter-disciplinary research direction. Using information diffusion over social networks as an example, it provides a thorough investigation of the impact of user behavior on the network evolution process, and demonstrates how such understanding can help improve network performance.

Note that there are also other works on swarm intelligence and evolutionary algorithms that reviews recent research on complex networks, and analyze the relations between the dynamics of evolutionary algorithms, complex networks and coupled map lattices [11–13]. These works often consider homogeneous networks and assume that all agents/smart entities in the complex network have a common goal. Different from these works, we address new challenges in crowd intelligence networks, consider the existence of irrational and/or malicious behavior, study the impact of user heterogeneity on evolutionary dynamics, and analyze the evolution of correlated events.

This book targets graduate students and researchers from different disciplines, including but not limited to data science, networking, signal processing, complex systems, and economics. It aims to encourage researchers in related research fields to explore many untouched territories along this direction, and ultimately to design crowd networks with efficient, effective, and reliable services.

References

1. D. Helbing, Globally networked risks and how to respond. Nature **497**, 51–59 (2013)
2. D. Pierson, Japan radiation fears speark panic salt-buying in China, *Los Angeles Times*, March 18, 2011, [Online]. Available: http://articles.latimes.com/2011/mar/18/world/la-fg-china-iodine-salt-20110318
3. COVID-19 and social impact, *Forbes*, April 10, 2020, [Online]. Available: https://www.forbes.com/sites/sorensonimpact/2020/04/10/covid-19-and-social-impact/#3a662623546b
4. Wikipedia, Social network. [Online]. Available: http://en.wikipedia.org/wiki/Social_network
5. K. Hao, T. Basu, The coronavirus is the first true social-media 'infodemic', *MIT Technology Review*, Feb. 12, 2020, [Online]. Available: https://www.technologyreview.com/2020/02/12/844851/the-coronavirus-is-the-first-true-social-media-infodemic
6. R. Pastor-Satorras, A. Vespignani, Epidemic spreading in scale-free networks. Phys. Rev. Lett. **86**(14), 3200–3203 (2001)
7. D. Gruhl, R. Guha, D. Liben-Nowell, A. Tomkins, Information diffusion through blogspace, in *Proceedings of 13rd International Conference on World Wide Web (WWW)* (2004), pp. 491–501
8. Evolutionary game theory, Wikipedia. [Online]. Available: https://en.wikipedia.org/wiki/Evolutionary_game_theory
9. D. Friedman, B. Sinervo, *Evolutionary Games in Natural, Social and Virtual Worlds* (Oxford University Press, Oxford, 2016)
10. J. Tanimoto, *Evolutionary Games with Sociophysics: Analysis of Traffic Flow and Epidemics* (Springer, Berlin, 2018)
11. I. Zelinka, G. Chen, *Evolutionary Algorithms Swarm Dynamics and Complex Networks* (Springer, Berlin, 2018)
12. T.W. Malone, M.S. Bernstein, *Handbook of Collective Intelligence* (MIT Press, 2015)
13. D. Miorandi, V. Maltese, M.N. Rovatsos, J. Stewart, *Social Collective Intelligence* (Springer, Berlin, 2014)

Chapter 2
Evolutionary Dynamics with Rational Users

Abstract Social networks have become ubiquitous in our daily life, and people are now used to interacting and sharing information through social networks. Understanding the mechanisms of tremendous information propagation over social networks is critical to various applications such as online advertisement and rumour control. In this chapter, we focus on the evolutionary game-theoretic model for information diffusion among rational users in social network, and analyze evolutionary dynamics under several different scenarios. By applying graphical evolutionary game theory (EGT) to information diffusion, we could predict every small change in the process, get the detailed dynamics and finally foretell the stable states.

Keywords Evolutionary game theory · Evolutionary dynamics · ESS · Information diffusion · Social networks

2.1 Introduction

The social network plays a profound role as a medium for the diffusion of information, ideas, and influence among its users. With the rapid development of the Internet and mobile technologies, today's social networks are of extremely large scale, and the information size on the social networks is becoming even tremendous-scale. For example, only in 2014, about 500 millions of tweets are sent from Twitter every day [1] while around 300 thousand statuses are updated every minute on Facebook [2].

The information spread through social networks is diverse, e.g., when a total victory is achieved in a sport match, some political opinions are declared by a party or politics, titbits or rumours about a superstar are exposed to the public, advertisements of various products are posted. All these information will go through the process of generation, dissemination and disappearance, the most important of which is the diffusion of information. Each piece of information may last for a long time, i.e. become very popular, or disappear quickly with few impacts. The diffusion dynamics or the popularity of the information is determined by complicated interactions and decision-making of many users. According to many factors such as the preferences

Y. Chen and H. V. Zhao, *Behavior and Evolutionary Dynamics in Crowd Networks*, Lecture Notes in Social Networks, https://doi.org/10.1007/978-981-15-7160-2_2

of different people, the behavior of surrounding neighbors, and the authenticity of the information, users would choose whether to spread information. For instance, when a consumer sees an advertisement for a new product, he/she will decide whether to trust or resist based on the comments of friends, the reputation of the sender and manufacturer, or his/her initial impression. Of course, if the consumer is a fan of the product, he/she tends to share this information on social networks, which strengthens the connection between the consumer and information sender. However, if the consumer isn't interested in it at all, the information spreader may be regarded as worthless and thus the connection would be cut. In practice, the mechanisms in the process of information diffusion should be explored in depth.

In the literature, there are numerous works on the information diffusion over social networks. Based on the research object, existing works can be classified into three categories: diffusion characteristics analysis, diffusion dynamics analysis and diffusion stability analysis. As for the first category, in [3] authors discussed how to extract the most influential nodes on a large-scale social network. A network growth model that can produce networks with necessary features for analysis was proposed, and how each feature affects information diffusion was also analyzed in [4]. In addition, many methods were proposed to mine top-k influential nodes in mobile social networks, e.g., a community-based greedy algorithm in [5], the Shapley value-based Influential Nodes algorithm in [6], and content-based improved greedy algorithm in [7] which reduced the total amount of computations. The second type is to analyze the dynamic diffusion process of different networks with different mathematical models [8–12]. In [8], Damon studied how a social network affected the spread of behavior and investigated the effects of network structure on users' behavior diffusion. The role of social networks in general information diffusion was studied through an experimental approach in [9]. With the prevalence of online social networks, there are many empirical analysis using large-scale datasets, including predicting the speed and range of information diffusion on Twitter [10], modeling the global influence of a node on the rate of diffusion on Memetracker [11] and illustrating the statistical mechanics of rumor spreading on Facebook [12]. The third category paid attention to the stability and consequence of information diffusion [13–18]. Authors in [13] studied the conditions for information diffusion vanishing and information diffusion being persistent in social networks. Peng et al. used a mathematical model to under the scenario of multi-source news and validated its accuracy [14]. In [15, 16], how to restrain the private or contaminated information diffusion was studied through identifying the important information links and hubs. Authors studied the method of maximizing information diffusion in [17] by designing effective neighbors selection strategies, while in [18] approximation algorithms were proposed to realize the influence maximization.

From the perspective of adopted methods, works could also be divided into two categories: (i) using machine learning (ML) or data mining approaches to make inference and prediction; (ii) devising microscopic mechanisms to explain the information diffusion from the perspective of the individual users' interactions. Among the first category, authors predicted future diffusion by early diffusion data in [5] while the community structure was further exploited to improve the prediction performance of viral memes in [20]. Given the information diffusion data, authors proposed efficient algorithms to infer the underlying information diffusion network in [21–23]. A matrix factorization based predictive model was proposed and gradient descent was used to optimize objective function in [12]. In [13], Alsuwaidan et al. proposed a novel model based on a physical radiation energy transfer mechanism to predict the diffusion graph of a certain contagion. By a rank-learning based data-driven approach, authors in [14] studied diffusion of preference on social networks. A K-center method was proposed in [15] to realize multi-source identification of information diffusion and the corresponding infection regions in general networks. In [16], authors analyzed various heuristic based influence maximization techniques and proposed a machine learning based approach to find the spread of information in the network. In the second category, a dynamic activeness model was proposed in [18] to study the problem of predicting dynamic trends according to each user's activeness. Assuming that each user responded best to the population's strategies, Morris studied the conditions for global contagion of behaviors [25]. Based on the correlation, authors in [19] proposed a probabilistic model to estimate the probability of users' adopting the naive Bayes classifier. In [20, 21], a game-theoretic framework was proposed for the study of competition between firms who aimed to maximize adoption of their products among consumers in a social network. The authors in [23–25] studied information spreading by defining different objective functions for each user and then settling corresponding minimization or maximization problem.

Recently, some EGT frameworks are proposed to model the users' interactions during the information diffusion process. Authors in [26, 27] proposed an evolutionary game-theoretic framework to model the dynamic information diffusion process among nodes in social networks, where the authors in [26] paid more attention to the final stable state, while in [27] the emphasis was on the evolutionary dynamics. Cao et al. in [28] then extended the analysis of the information diffusion process to the heterogeneous social networks where nodes can have different types. It was found in these works that the dynamics derived under the evolutionary game framework fit the real-world information diffusion dynamics well and could even make predictions on the future diffusion dynamics, suggesting a suitable and tractable paradigm for analyzing the information diffusion. Therefore, in this chapter, we introduce information diffusion analysis based on evolutionary game-theoretic model, by summarizing the results in [26–28], with the assumption that users in the social network are all rational.

2.2 Graphical Evolutionary Game Formulation

Due to the intricate connections between users in the social network structure, we combine the graphical presentation of the network with EGT to analyze the information diffusion as it makes problem more straightforward. As shown in Fig. 2.1, the social network could be denoted as a graph. In heterogeneous social networks, nodes with different colors represent different types of users. Or if users are treated as homogeneous individuals, there is no difference between nodes with different

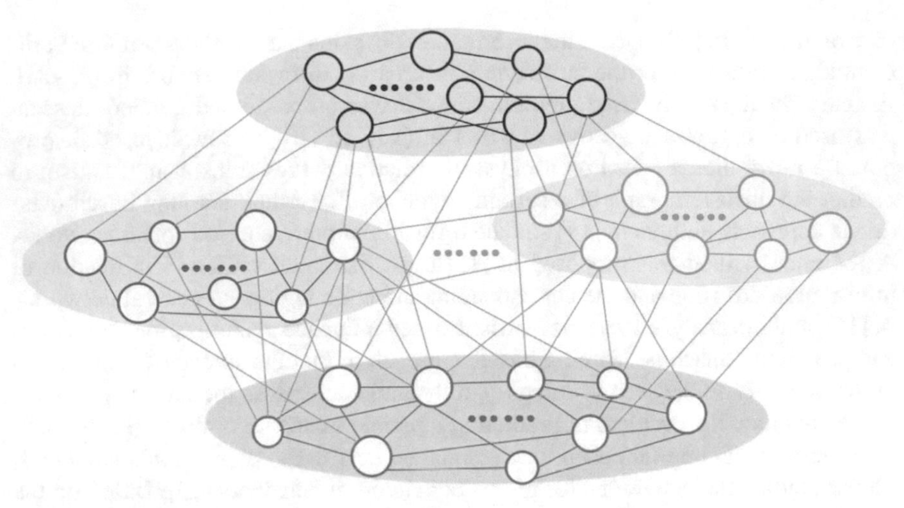

Fig. 2.1 An illustration of the social network

colors in the analysis. In the information diffusion process, each user has two possible strategies towards received information: forwarding indicated as S_f and not forwarding indicated as S_n, and strategies are known for users with connections under all scenarios. Thus for the center node with a certain amount of neighbors, the numbers of neighbor nodes adopting S_f and S_n are certainly available. When two connected users interact with each other with their own strategies, both sides would get instant payoff, which equals to the benefit of interaction and could be obtained from a predefined payoff matrix:

$$\begin{array}{cc} & \begin{array}{cc} S_f & S_n \end{array} \\ \begin{array}{c} S_f \\ S_n \end{array} & \begin{pmatrix} u_{ff} & u_{fn} \\ u_{nf} & u_{nn} \end{pmatrix} \end{array} \tag{2.1}$$

Based on the payoff, the fitness of every user could be calculated, involving baseline fitness B, payoffs U and selection intensity α as

$$\Psi = (1 - \alpha) \cdot B + \alpha \cdot U. \tag{2.2}$$

Baseline fitness means the inherent property of a player, e.g., the user's interests on the released news. In all frameworks, the baseline fitness is normalized as one. Payoffs are determined by the payoff matrix and the graph structure. And selection intensity is the relative contribution of the game to fitness. When $\alpha \to 0$, it indicates the limit of weak selection [29, 30], while $\alpha \to 1$ denotes strong selection, where fitness equals payoffs. Here assuming that the weak selection is adopted thus selection intensity is a small value.

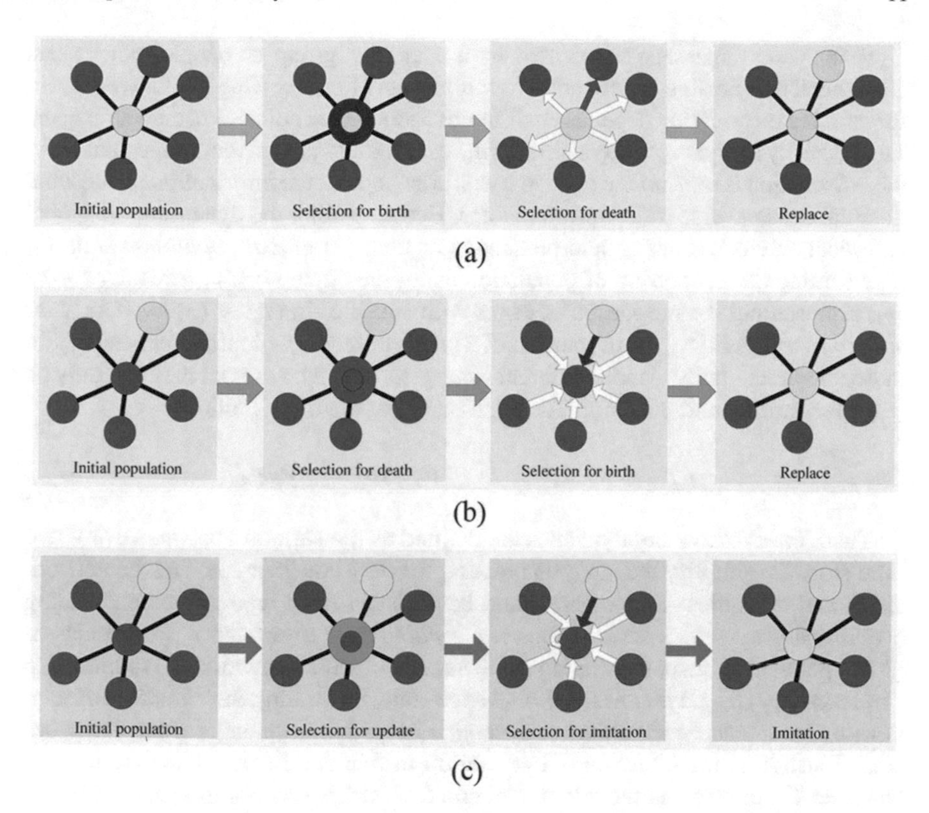

Fig. 2.2 Three different update rules, where death selections are shown in dark blue and birth selections are shown in red. **a** BD update rule.; **b** DB update rule.; **c** IM update rule

With strategies and fitness, we introduce three most typical and prevalent rules for strategy update, namely birth-death (BD), death-birth (DB) and imitation (IM) [31] (shown in Fig. 2.2).

1. BD updating rule: a player is chosen for reproduction with the probability being proportional to fitness (Birth process). Then, the chosen player's strategy replaces one neighbor's strategy with uniform probability (Death process).
2. DB updating rule: a random player is chosen to abandon his/her current strategy (Death process). Then, the chosen player adopts one of his/her neighbors' strategies with the probability being proportional to their fitness (Birth process).
3. IM updating rule: each player either adopts the strategy of one neighbor or remains his/her current strategy, with the probability being proportional to fitness.

These strategy update rules are from the evolutionary biology field and used to model the resident/mutant evolution process. According to [27], the analyses of these three update rules are equivalent under weak selection over the network with sufficiently large degree.

When a new message is released by a user or a group of people, other users are expected to be able to forward it to more users after receiving it. But whether to forward the information depends on different users' own choices, i.e. their strategies. Therefore, by analyzing the dynamics of the user's strategies, we could speculate how the information is transmitted to other users, how popular the information is, and what the final consequence of the information is. Before getting the dynamics, the global and local network states should be defined at first: (1) *global population state* p_i, representing the proportion of population using the strategy; (2) *global edge state* p_{ij}, representing the proportion of edges with specific strategies; (3) *local network state* $p_{i|j}$, representing the proportion of a user's neighbors adopting the strategy. In these notations i and j denote different strategies, and in the model they can only be f for forwarding and n for not. The relationship of them is as follows:

$$p_{f|f} = p_{ff}/p_f, \quad (1 - p_{f|f})p_f = p_{f|n}p_n. \tag{2.3}$$

Then, the evolutionary dynamics are defined as the variation between every two time slots. To simplify the analysis process, we only consider some of the network states and dynamics since others could be deduced from them, so corresponding evolutionary dynamics are as follows: (1) *population dynamics* $\dot{p}_f$: dynamics of global population state, illustrating the dynamics of whole network; (2) *relationship dynamics* $\dot{p}_{ff}$, $\dot{p}_{nn}$: dynamics of global edge state, illustrating the dynamics of relationship among users; (3) *influence dynamics* $\dot{p}_{f|f}$, $\dot{p}_{f|n}$: dynamics of local network state, illustrating the influence of one user on his/her neighbors. Note that the time has been discretized and the duration of one time slot is set appropriately according to different conditions. In each time slot, users would observe the strategies of neighbors in the population, then they will decide on their strategies in the next time slot, i.e., retain the current strategy or deviate from the current to the other strategy based on the observed information of neighbors' strategies and calculated fitness. Thus, as the users' strategies update slot by slot, the dynamics and network state also keep changing slot by slot.

As the information dissemination process progresses, the entire network would gradually reach an evolutionary stable state (ESS), and all users tend to adopt their optimal stable strategies [4]. Generally, when the entire population is adopting the optimal strategy, the fitness of a small group of "invaders" using any alternative strategy should be strictly lower than most users, and they eventually die with a high probability. To get final ESSs, the most common method is to find the stable points of different kinds of dynamics.

In the following, with different network structure and scenario setting, we will give the analysis of homogeneous network and heterogeneous network respectively. No matter which kind of scenario is, the basic framework can be summarized as follows. First of all, we need to predict the certain amount of neighbors adopting S_f and S_n in every time slot. On account of the assumption that the social network is large enough, it's reasonable to use global population state p_i or local network state $p_{i|j}$ as the probability that the center user encounters a neighbor adopting strategy i. Thus, with the total number of neighbor nodes, the distribution of the number of neighbor nodes using every strategy can be obtained. Then based on the numbers, predefined baseline fitness, payoffs and selection intensity, we could get each user's fitness, which is proportional to the probability of a user being chosen to reproduction or updating his/her strategy as stated before in strategy updating rules. With the probability of strategy change, the probability for the increase or decrease of evolutionary states could also be known, and this is exactly the evolutionary dynamics that we look for. By setting the evolutionary dynamics as zero the final ESS under different conditions could finally be computed. Based on the derived formula of evolutionary dynamics and final ESS, we are able to predict the evolutionary process in every time slot, and foretell the stable states of the information diffusion, e.g. how many users sent this piece of message in the social network. Next we seek out several different network structures to illustrate the framework.

2.3 Analysis of Homogeneous Network

2.3.1 Diffusion Dynamics and ESSs Over Uniform Degree Networks

In the uniform scenario, a N-user homogeneous social network with the same degree k for all users is considered. More specifically, in homogeneous network with uniform degree, there is no difference between all users so they could be regarded as the same type of whole, and every user has the same number of neighbor nodes k. After discussing the uniform network, we would have a deeper understanding of the complex information diffusion problem, and the derivation and results such as fitness calculation and dynamics derivations could also be applied in the non-uniform degree case.

Assuming that when the new information is released, whether users forward it or not follows the BD strategy update rule. Under the weak selection, the dynamics of three types could be derived as [27]:

Relationship Dynamics over Uniform Degree Networks

$$\dot{p}_{ff} = \frac{Np_f(1-p_{f|f})\left[1+(k-1)p_{f|n}\right] - N(1-p_f)p_{f|n}(k-1)p_{f|f}}{Nk/2}, \tag{2.4}$$

$$\dot{p}_{nn} = \frac{N(1-p_f)\left[1+(k-1)p_{n|f}\right] - Np_f(1-p_{f|f})(k-1)p_{n|n}}{Nk/2}. \tag{2.5}$$

Influence Dynamics over Uniform Degree Networks

$$\dot{p}_{f|f} = \frac{\dot{p}_{ff}}{p_f} = \frac{2}{k}\Big\{1+(k-1)\big[p_{f|f}p_{f|f}+p_{f|n}\big(1-p_{f|f}\big)\big]-kp_{f|f}\Big\}, \tag{2.6}$$

$$\dot{p}_{f|n}=1-\frac{\dot{p}_{nn}}{1-p_f}=\frac{2}{k}\Big\{(k-1)\big[p_{f|f}p_{f|n}+p_{f|n}(1-p_{f|n})\big]-kp_{f|n}\Big\}. \tag{2.7}$$

Population Dynamics over Uniform Degree Networks

$$\dot{p}_f = \frac{\alpha(k-2)}{(k-1)}p_f(1-p_f)\left(a_1p_f+b_1\right)+o(\alpha^2), \tag{2.8}$$

$$\text{where} \begin{cases} a_1 = (k-2)(u_{ff}-2u_{fn}+u_{nn}), \\ b_1 = u_{ff}+(k-2)u_{fn}-(k-1)u_{nn}. \end{cases} \tag{2.9}$$

Then by setting dynamics to zero, ESSs, i.e. network stable states, could be written in a closed form as [26, 27]:

Local Network Stable State over Uniform Degree Networks

$$p^*_{f|n} = \frac{(k-2)p_f}{k-1}, \quad p^*_{f|f} = \frac{(k-2)p_f+1}{k-1}. \tag{2.10}$$

Global Population Stable State over Uniform Degree Networks

$$p^*_f = \begin{cases} 1, & \text{if } \ u_{ff} > u_{fn} > u_{nn}, \\ 0, & \text{if } \ u_{nn} > u_{fn} > u_{ff}, \\ \frac{b_1}{a_1} = \frac{u_{ff}+(k-2)u_{fn}-(k-1)u_{nn}}{(k-2)(2u_{fn}-u_{ff}-u_{nn})}, & \text{else.} \end{cases} \tag{2.11}$$

Generally, dynamics and ESSs for the global population state can best reflect the condition for information diffusion, so we focus on p_f and p^*_f. It could be observed from the formula in (2.8) and (2.9) that the population dynamics only rely on the initial population state, the values of payoff matrix and the degree of the network, regardless of the network scale information. Therefore, the population dynamics of information diffusion over uniform degree networks show the scale-free property. Moreover, the formula reveals that the dynamic is an increasing function of the payoff u_{ff}, which is consistent with the intuition that the percentage of people forwarding the information would be higher if higher payoff can be obtained by forwarding the information. On the contrary, if not forwarding the information is more favorable, the percentage will be lower, which is the reason that population dynamics p_f is a decreasing function of the payoff u_{nn}. As for the relationship dynamics and influence dynamics, they are functions involving themselves as well as the population dynamics, which are more sophisticated.

From (2.11), we could see that there are three possible ESSs, which are respectively 1, 0 and a value between 0 and 1. Zero and one are two extreme stable states, signifying that no user forwards and all users forward the information, respectively. When the population state is 1, it means that users choose to forward the information to obtain the largest gain, while the non-forwarding gain is the smallest. In social networks, this corresponds to the situation where the posted information is an extremely hot topic and reposting it could attract more attention. Reversely, when the population stable state is 0 the situation is just the opposite. This corresponds to the situation where the published information is useless or a negative advertisement, and forwarding it would only generate unnecessary costs. As for the third ESS, there are two other cases, one of which is $u_{fn} > u_{nn}, u_{fn} > u_{ff}$. In this case, unilateral forwarding is beneficial than no forwarding or simultaneous forwarding, which is corresponding to the scenario where both users can only receive limited returns but increase the cost of both parties for forwarding information. For example, when the information is not the mainstream topic, such as the news about a punk musician, it is supposed to be spread among people with similar interests. In the other case $u_{nn} > u_{fn}, u_{ff} > u_{fn}$, the payoff configuration is equivalent to that of the coordination game, where the gains of two parties with the same action are greater than the gains of the opposite action. An example is that the information is politically sensitive and its reality is not guaranteed, forwarding which may be attractive but also incur potential misleading cost.

2.3.2 *Diffusion Dynamics and ESSs Over Non-uniform Degree Networks*

In the non-uniform scenario, we consider a N-user social network based on a graph whose degree exhibits a specific distribution $\lambda(k)$. The meaning of this distribution is that when a user is randomly selected on the network, the probability that the chosen user has k neighbors is $\lambda(k)$. So the average degree of the network is

$$\overline{k} = \sum_{k=0}^{+\infty} \lambda(k)k. \tag{2.12}$$

Here we don't consider the degree correlation, or in other words, the degrees of all users are independent. Hence, when new information is released, all users will voluntarily update their own strategies. According to the BD strategy update rule, there are two types of users: the selected user and the replaced neighbor. Obviously, the degree of chosen user follows the distribution $\lambda(k)$. But for the replaced neighbor, its degree doesn't obey $\lambda(k)$ because if a pair is selected by random, the degree distribution of the user on the specific pair is $\frac{k\lambda(k)}{\sum_{k=0}^{+\infty} k\lambda(k)}$ rather than $\lambda(k)$ [32]. By taking expectations with respect to different users' degree, the population dynamics of information diffusion over non-uniform degree networks could be derived as [27] Influence Dynamics over Non-Uniform Degree Networks

$$\dot{p}_{f|f} = \frac{2}{\overline{k}}\left\{1+\left(\frac{\overline{k^2}}{\overline{k}}-1\right)\left[p_{f|f}p_{f|f}+p_{f|n}(1-p_{f|f})\right]-\frac{\overline{k^2}}{\overline{k}}p_{f|f}\right\}, \tag{2.13}$$

$$\dot{p}_{f|n} = \frac{2}{\overline{k}}\left\{\left(\frac{\overline{k^2}}{\overline{k}} - 1\right)\left[p_{f|f}p_{f|n} + p_{f|n}(1 - p_{f|n})\right]-\frac{\overline{k^2}}{\overline{k}}p_{f|n}\right\}. \tag{2.14}$$

Population Dynamics over Non-Uniform Degree Networks

$$\dot{p}_f = \frac{\alpha(\overline{k}-1)(\overline{k^2}-2\overline{k})}{(\overline{k^2}-\overline{k})^2}p_f(1-p_f)\left(a_2p_f+b_2\right)+o(\alpha^2), \tag{2.15}$$

$$\text{where}\begin{cases} a_2 = (\overline{k^2} - 2\overline{k})(u_{ff} - 2u_{fn} + u_{nn}),\\ b_2 = \overline{k}u_{ff} + (\overline{k^2} - 2\overline{k})u_{fn} - (\overline{k^2} - \overline{k})u_{nn}. \end{cases} \tag{2.16}$$

The corresponding ESSs are as follows:

Local Network Stable State over Non-Uniform Degree Networks

$$x^*_{f|f} = \frac{(\overline{k^2} - 2\overline{k})x_f + \overline{k}}{\overline{k^2} - \overline{k}}, \quad x^*_{f|n} = \frac{(\overline{k^2} - 2\overline{k})x_f}{\overline{k^2} - \overline{k}}. \tag{2.17}$$

Global Population Stable State over Non-Uniform Degree Networks

$$p^*_f = \begin{cases} 1, & \text{if } u_{ff} > u_{fn} > u_{nn},\\ 0, & \text{if } u_{nn} > u_{fn} > u_{ff},\\ \dfrac{\left(\frac{\overline{k}^2}{\overline{k}-2}\right)(u_{fn}-u_{nn})+(u_{ff}-u_{nn})}{\left(\frac{\overline{k}^2}{\overline{k}-2}\right)(2u_{fn}-u_{ff}-u_{nn})}, & \text{else.} \end{cases} \tag{2.18}$$

From these formulae, we could see that the procedure and conclusions are similar to those in uniform degree network, and the biggest difference is the introduction of the probability distribution of degrees, so that the expectation and variance appear.

2.4 Analysis of Heterogeneous Network

2.4.1 Theoretical Analysis for the Unknown User Type Model

In view of different habits and interests of individuals, the social network could be modeled as a heterogeneous network, consisting of various types of users. For example, if there is a group of people who are all sports fans, then they belong to the same type and will consider forwarding sports-related information, while some people who are music lovers are in the same type. Consequently, for one piece of information, different types of users have different payoff matrices. In real-world social networks, users often do not know the types of their neighbors when they get in touch at first and not familiar with each other, thus in this subsection we present the unknown user type model over N-user M-type social network where the user type is private information that is unknown to others. In such a case, payoffs change with only one argument, i.e., the type of center user.

In this model, DB strategy update rule is adopted. By analyzing each type of user separately, the expected population dynamics for each type i could be obtained as [28]

$$\begin{aligned}\dot{p}_f(i) &= \frac{1}{N}p_f - \frac{1}{N}p_f(i) + \frac{\alpha}{N}p_f(p_f-1)\times\\ &\left[\Delta(i)\left(\left(\overline{k}-3+2\overline{k^{-1}}\right)p_f+1-\overline{k^{-1}}\right)+\Delta_n(i)(\overline{k}-1)\right],\end{aligned} \tag{2.19}$$

where $\overline{k}$ and $\overline{k^{-1}}$ denote the expectation of the network degree k and k^{-1} respectively, and $\Delta(i)$, $\Delta_n(i)$ are the combination of payoffs denoted as $\Delta(i) = 2u_{fn}(i) - u_{ff}(i) - u_{nn}(i)$, $\Delta_n(i) = u_{nn}(i) - u_{fn}(i)$ respectively. Then by proportional merging, the global evolutionary population dynamics could be obtained as [28]

$$\begin{aligned}\dot{p}_f &= \sum_{i=1}^{M} q(i)\dot{p}_f(i)\\ &= \frac{\alpha}{N}p_f(p_f-1)\left[\overline{\Delta}\left(\left(\overline{k}-3+2\overline{k^{-1}}\right)p_f+1-\overline{k^{-1}}\right)+\overline{\Delta}_n(\overline{k}-1)\right],\end{aligned} \tag{2.20}$$

where $\overline{\Delta} = \sum_{i=1}^{M} q(i)\Delta(i)$ and $\overline{\Delta}_n = \sum_{i=1}^{M} q(i)\Delta_n(i)$. $q(i)$ means the proportion of type-i users. It can be observed from (2.19) that the dynamic of each type not only consists of its current population state but also the global population state, signifying that nodes are affected not only by those with the same type but also by all other nodes. Additionally, comparing (2.20) with the population dynamics of homogeneous social network given in (2.8) and (2.15), we note that the global population dynamics $\dot{p}_f$ evolve as if the network is homogeneous with corresponding payoff matrix being the weighted average (with weights $q(i)$) of those among all the types. Similarly, by setting the dynamics as zero, ESSs of global state and local state are derived as [28]

$$p_f^* = \begin{cases} 0, & \text{if } \overline{u}_{nn} > \overline{u}_{fn},\\ 1, & \text{if } \overline{u}_{ff} > \overline{u}_{fn},\\ \dfrac{\overline{\Delta}_n(1-\overline{k})+\overline{\Delta}(\overline{k^{-1}}-1)}{\overline{\Delta}(\overline{k}-3+2\overline{k^{-1}})}, & \text{if } \max\{\overline{u}_{ff},\overline{u}_{nn}\} < \overline{u}_{fn},\end{cases} \tag{2.21}$$

$$p_f^*(i) = p_f^* + \alpha p_f^*(p_f^*-1)\left[\Delta(i)\left(\left(\overline{k}-3+2\overline{k^{-1}}\right)p_f^*+1-\overline{k^{-1}}\right)+\Delta_n(i)(\overline{k}-1)\right], \tag{2.22}$$

where $\overline{u}_{ff} = \sum_{i=1}^{M} q(i)u_{ff}(i)$ and $\overline{u}_{fn}, \overline{u}_{nn}$ are similarly defined. According to (2.22), the stable state of each type is related to the stable state of whole population. Meanwhile, it could be seen from (2.19) that when p_f is stable, all the type specific population dynamics $p_f(i)$ will also converge to their respective ESSs.

2.4.2 Theoretical Analysis for the Known User Type Model

Through repeated interactions, users may somehow know the type of their neighbors. For instance, when a user notices that one of his friends posts news about football matches regularly, he may be aware that the friend is a football fan. This condition could be modeled as known user type where the users' types are publicly known information. In this subsection we consider the known user type model over a N-user social network with degree k. In this case, payoffs change with two arguments, which are the strategies of two sides. The center user will treat his/her neighbors in different ways depending on their types when he/she is chosen to decide on whether to change the current strategy. Since a user's type and strategy affect its neighbors' payoffs, they would also influence neighbors' strategies. Therefore, not only the global population information but also the edge information including relationship state (global edge state) and influence state (local network state) are necessary in the analysis, so as to fully characterize the network state. The dynamics of different types are shown as follows [28]:

Population Dynamics

$$\begin{aligned}\dot{p}_f(i) = &\frac{\alpha\overline{k}}{N} p_f(i) p_{n|f}(i,i)(p_{n|n}(i,i) + p_{f|f}(i,i)) \\ &\times \sum_{j=1}^{M} q(j)[p_{f|f}(j,i)u_{ff}(i,j) + p_{n|f}(j,i)u_{fn}(i,j) \\ &- p_{f|n}(j,i)u_{nf}(i,j) - p_{n|n}(j,i)u_{nn}(i,j)],\end{aligned} \tag{2.23}$$

Relationship Dynamics

$$\begin{aligned}\dot{p}_{ff}(i,l) &= \frac{2}{N} q(i)q(l)p_f(i)(1 - p_{f|f}(i,i)) \times \left[\frac{p_f(l)}{p_n(i)}(1 - p_{f|f}(i,l)) - p_{f|f}(l,i)\right] \\ &+ \frac{2}{N} q(i)q(l)p_f(l)(1 - p_{f|f}(l,l)) \times \left[\frac{p_f(i)}{p_n(l)}(1 - p_{f|f}(l,i)) - p_{f|f}(i,l)\right],\end{aligned} \tag{2.24}$$

$$\dot{p}_{ff}(i,i) = \frac{2}{Np_n(i)} q^2(i)p_f(i)(1 - p_{f|f}(i,i))(p_f(i) - p_{f|f}(i,i)). \tag{2.25}$$

Influence Dynamics

$$\dot{p}_{f|f}(l,i) = \frac{1}{N}(p_f(l) - p_{f|f}(l,i)) \times \left[\frac{1 - p_{f|f}(i,i)}{p_n(i)} + \frac{1 - p_{f|f}(l,l)}{p_n(l)}\right], \forall l \neq i. \tag{2.26}$$

$$\dot{p}_{f|f}(i,i) = \frac{2}{Np_n(i)}(1 - p_{f|f}(i,i))(p_f(i) - p_{f|f}(i,i)), \forall i. \tag{2.27}$$

In these dynamics, i, j, l represent different types of users, and other denotations are the same as those in Sect. 2.4.1. From (2.23), it could be deduced that the population dynamics $\dot{p}_f(\cdot)$ evolves at the speed of $O(\alpha)$, while the relationship dynamics $\dot{p}_{ff}(\cdot,\cdot)$ and the influence dynamics $\dot{p}_{f|f}(\cdot,\cdot)$ evolve at the speed of $O(1)$ in (2.24), (2.25), (2.26) and (2.27). Due to the assumption that α is quite small, the conclusion is that the relationship dynamics and influence dynamics change at a much faster speed than population dynamics do. This implies we can choose a time window of appropriate length to keep the population dynamics basically unchanged, while the relationship dynamics and influence dynamics vary greatly. In this small time window, it has been proved that the influence state $p_{f|f}(\cdot,\cdot)$ will converge to the corresponding population state $p_f(\cdot)$, thus we can make the approximation that $p_{f|f}(l,i) = p_f(l), \forall l, i$. Accordingly, the population dynamics can be further simplified into the following form [28]:

$$\dot{p}_f(i) = \frac{\alpha\bar{k}}{N} p_f(i) p_n(i) \sum_{j=1}^{M} q(j)[p_f(j)(u_{ff}(i,j) - u_{nf}(i,j)) + p_n(j)(u_{fn}(i,j) - u_{nn}(i,j))]. \tag{2.28}$$

2.5 Simulations and Experiments

In this section, experiments are conducted to validate the theoretic derivations of four frameworks in Sects. 2.3 and 2.4, and the data used to perform the simulations consist of both synthetic ones and real-world data.

2.5.1 *Simulations Over Homogeneous Network*

First of all, in Fig. 2.3, conclusions are verified in three synthetic networks, each of which contains 1000 users:

- the uniform degree network;
- the Erdős-Rényi random network;
- the Barabási-Albert scale-free network.

The Erdős-Rényi random network and the Barabási-Albert scale-free network are all non-uniform degree networks. And four different payoff matrices are considered as:

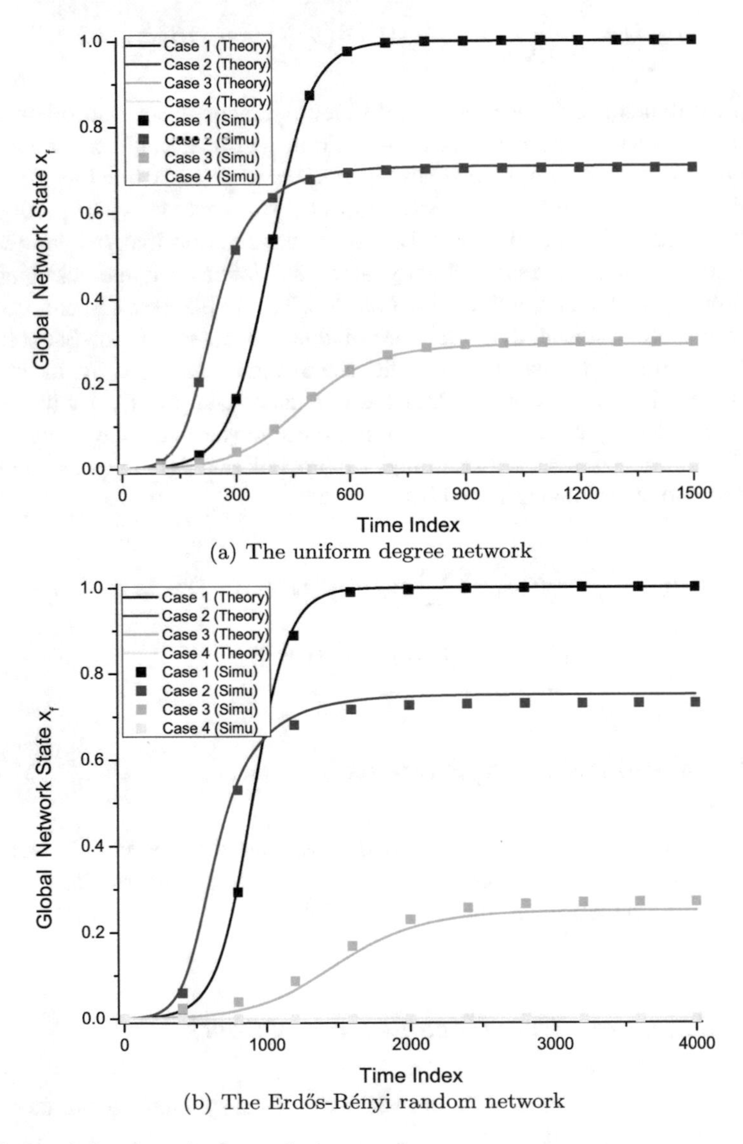

(a) The uniform degree network

(b) The Erdős-Rényi random network

Fig. 2.3 Population dynamics for synthetic networks

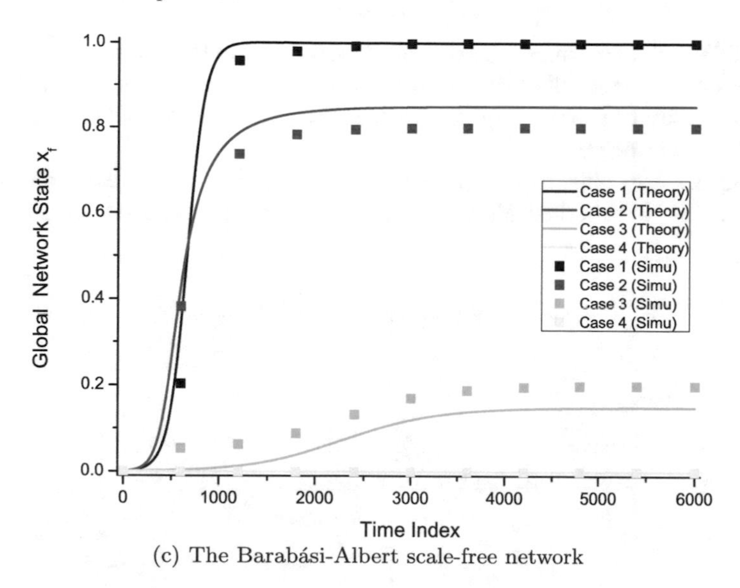

(c) The Barabási-Albert scale-free network

Fig. 2.3 (continued)

- Case 1: $u_{ff} = 0.8 > u_{fn} = 0.6 > u_{nn} = 0.4$;
- Case 2: $u_{fn} = 0.8 > u_{ff} = 0.6 > u_{nn} = 0.4$;
- Case 3: $u_{fn} = 0.8 > u_{nn} = 0.6 > u_{ff} = 0.4$;
- Case 4: $u_{nn} = 0.8 > u_{fn} = 0.6 > u_{ff} = 0.4$.

From Fig. 2.3, it could be seen that different settings of the payoff matrix can result in different dynamics of information diffusion, and all simulation results agree well with the theoretical results, which demonstrate the correctness of the conclusions in (2.8) and (2.15). In Fig. 2.3c, the reason for the gap of the Barabási-Albert scale-free network is that there is a weak dependence between the global network state and the network degree, which is ignored in the diffusion analysis. The results of global population stable state, i.e. ESS, for the uniform degree network, Erdős-Rényi random network and Barabási-Albert scale-free network under different average degrees and four payoff matrices are shown in Fig. 2.4a, b, c, respectively. The theoretical results are calculated from (2.11) and (2.18) directly, while the simulation results are obtained by emulating the information diffusion process over the generated network. Also we could see that the simulation results are consistent with the theoretical analysis, and the network degree variations have little influence on the ESSs. In Fig. 2.4c, there are small gaps for the Barabási-Albert scale-free network, and the reason is the same as before. Nevertheless, the gap is relatively small and indeed negligible.

Besides synthetic networks, simulations over real-world Facebook networks are also conducted to evaluate analysis over the non-uniform network. Figure 2.5 shows the simulation results of dynamics under different payoff matrices settings over the Facebook network, which contains totally 4039 users and 88234 edges and the average degree of which is about 40 [33]. And Fig 2.6 shows the results of ESSs under different payoff matrices over ten ego-networks in the data set. It can be seen that the simulation results match well with the theoretical results, while the small gaps are mainly due to the neglected dependence between the global network state and the network degree.

In Fig. 2.7, the comparison of proposed model and one of the existing data mining method in [34] is exhibited. Here several Twitter hashtags are used, which are from

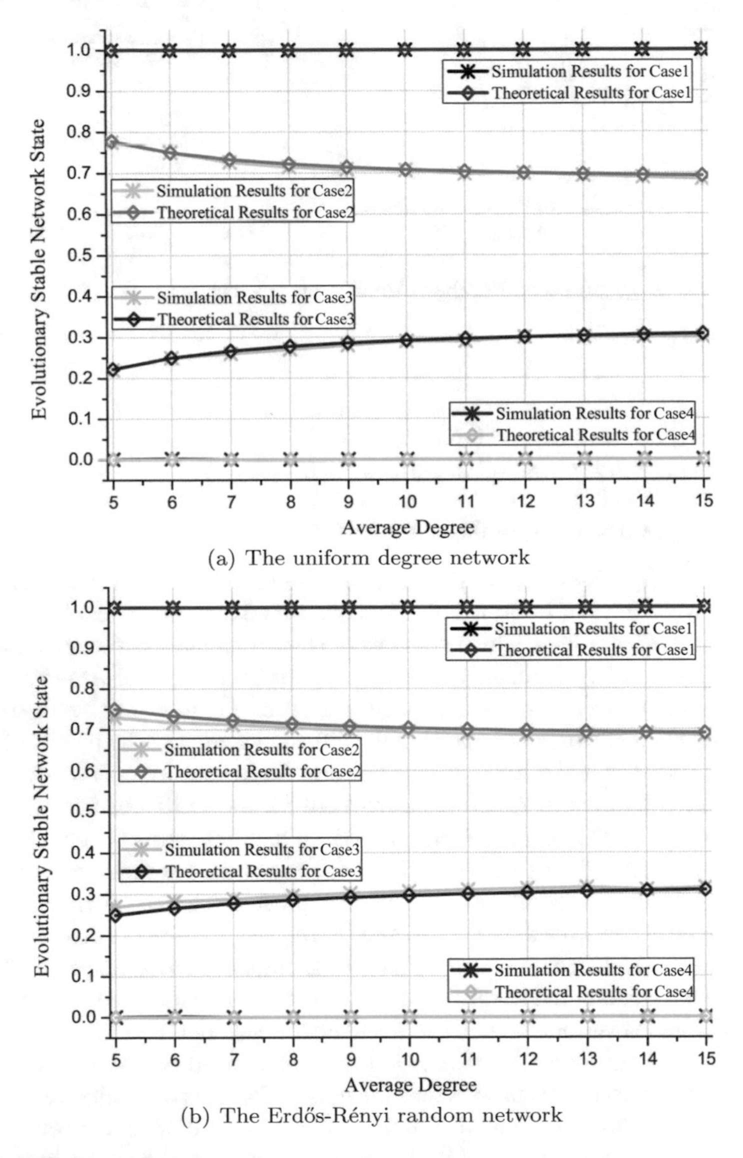

Fig. 2.4 ESSs for synthetic networks

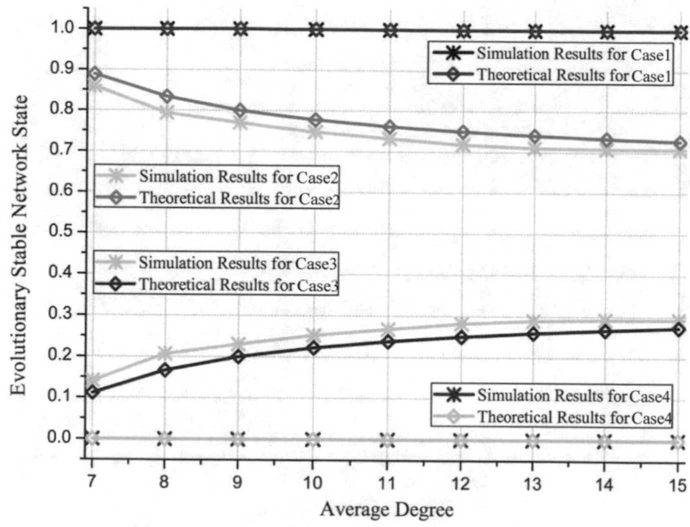

(c) The Barabási-Albert scale-free network

Fig. 2.4 (continued)

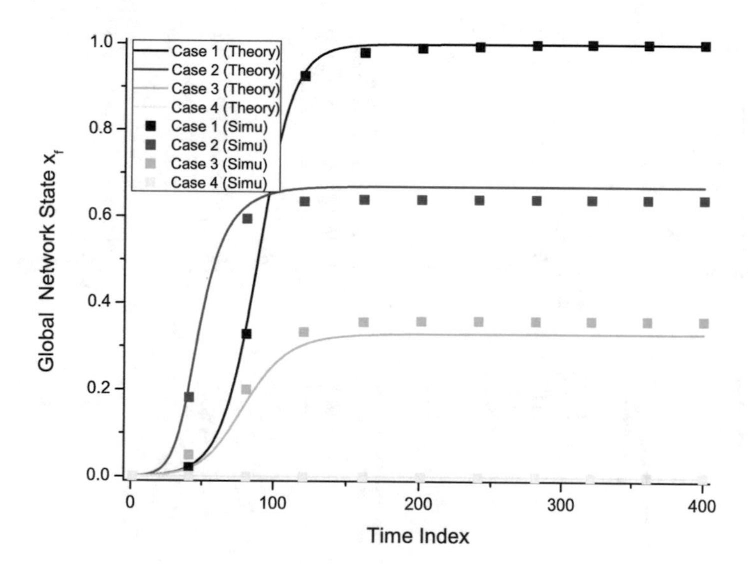

Fig. 2.5 Population dynamics for the real-world Facebook network

the Twitter hashtag data set in [33]. It contains the number of mention times per hour of 1000 Twitter hashtags with corresponding time series, which are the 1000 hashtags with highest total mention times among 6 million hashtags from Jun. to Dec. 2009. Different from simulations before, the payoff matrices are estimated using least squares method by fitting the curves in Fig. 2.7. The vertical axis is

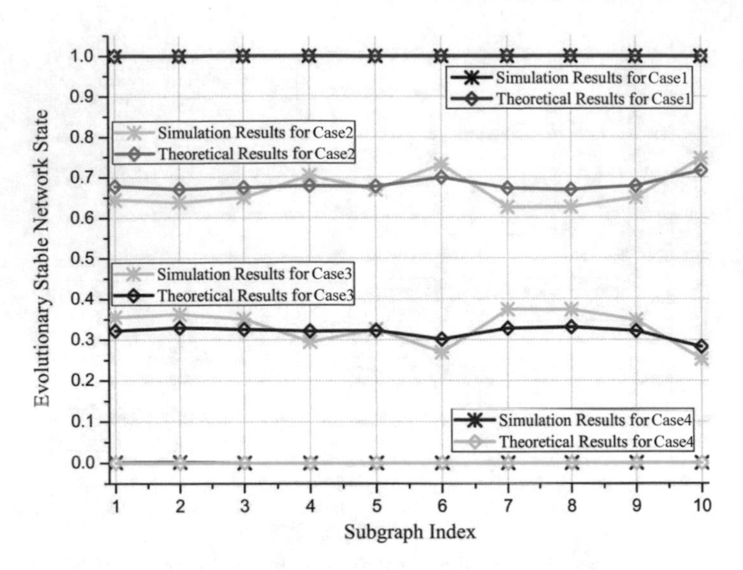

Fig. 2.6 ESSs for the real-world Facebook network

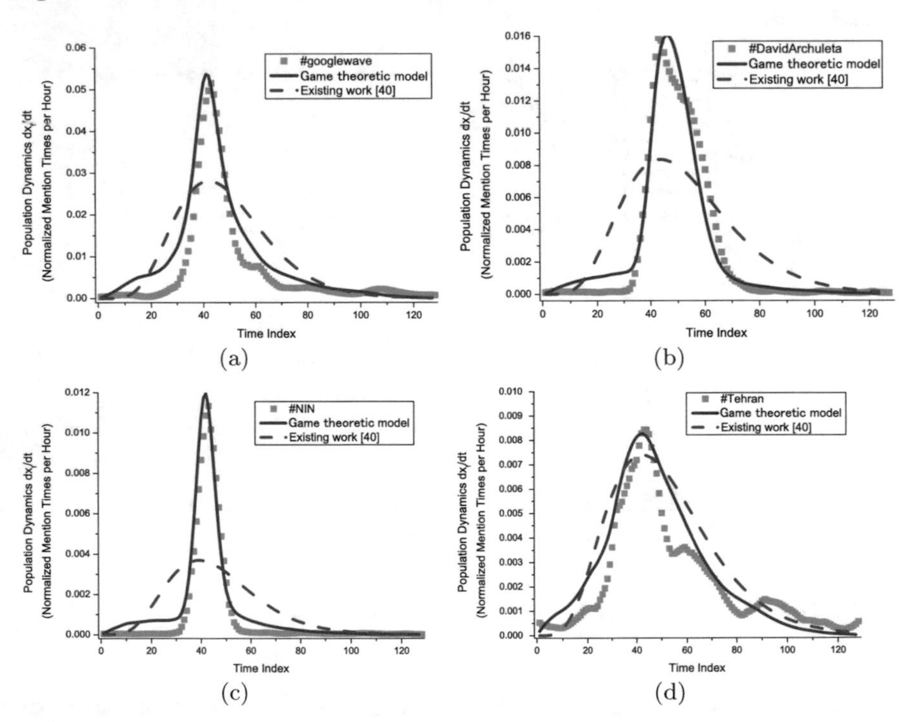

Fig. 2.7 Comparison with the existing work. **a** googlewave. **b** DavidArchuleta. **c** NIN. **d** Tehran

the dynamics, and the mention times of different hashtags per hour in the Twitter dataset are normalized within interval and denoted by solid grey square. From the figure, it could been observed that the graphical EGT model can fit the real-world information diffusion dynamics better than the data mining method in [34] since users' interactions and decision making behaviors are taken into account.

2.5.2 Simulations Over Heterogeneous Network

To validate the unknown user type model and known user type model for heterogeneous network, a constant degree network (k is a deterministic constant) consisting of two types of users is synthesized. The simulation results of unknown user type model under the payoff setting $u_{ff}(1) = 0.4, u_{fn}(1) = 0.6, u_{nn}(1) = 0.3, u_{ff}(2) = 0.2, u_{fn}(2) = 0.4, u_{nn}(2) = 0.5$ are shown in Fig. 2.8, in which the theoretic results are calculated from (2.19) and the simulated results are obtained by simulating the DB strategy update rule over the generated constant degree network. The theoretical dynamics fit the simulation dynamics well and ESSs are forecasted accurately. The average relative ESS error of the heterogeneous model is 3.54%. Moreover, thc simulation results of the evolutionary dynamics under another utility parameter setup $u_{ff}(1) = 0.5, u_{fn}(1) = 0.8, u_{nn}(1) = 0.1, u_{ff}(2) = 0.1, u_{fn}(2) = 0.5, u_{nn}(2) = 0.3$ are shown in Fig. 2.9 and it could also be seen that the simulated dynamics still match well with the theoretical ones. Simulation results for the known user type in Fig. 2.11 demonstrate that theoretical dynamics and the simulated dynamics match well and prove the correctness of theoretical analysis in Sect. 2.4.2. In Fig. 2.11, the evolutionary dynamics given by the theory of the unknown user type model are also plotted, but it doesn't conform to the simulated evolutionary dynamics under the known user type model, indicating that the theory of the known user type model is necessary.

Then we fit the theoretical dynamics for two models with the real data, i.e. Twitter hashtags from the dataset [35]. The dataset comprises sequences of adopters and timestamps for the observed hashtags, and four popular hashtags are selected from the whole to conduct experiments. We distinguish the users into two types to characterize the heterogeneity of the users, and the classification is based on the users' activity. Specifically, we calculate the number of hashtags each user has mentioned. Then the top 10% of users who post highest number of hashtags are categorized as Type-1 users, and the rest as Type-2 users. In Figs. 2.12 and 2.10, the real data denoted by marks are used to estimate the parameters, i.e., $\Delta(i)$ and $\Delta_n(i)$, and then the theoretical dynamics based on the estimated parameters are computed, which are denoted by solid line and fit well with real data (marks). It validates the effectiveness of considering the individuals' interactions, and also suggests the heterogeneous behavior dynamics of online users are consequences of their heterogeneous payoff structures.

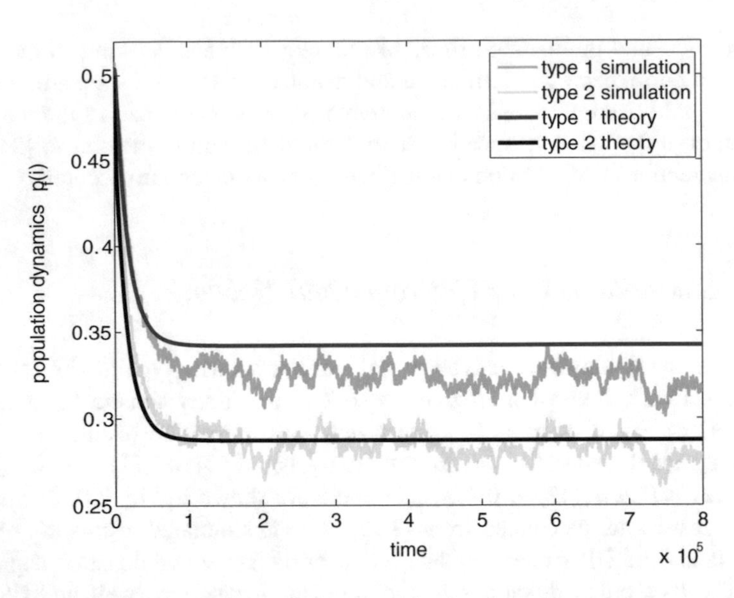

Fig. 2.8 Simulation results of the evolution dynamics for the unknown user type model with $u_{ff}(1) = 0.4$, $u_{fn}(1) = 0.6$, $u_{nn}(1) = 0.3$, $u_{ff}(2) = 0.2$, $u_{fn}(2) = 0.4$, $u_{nn}(2) = 0.5$

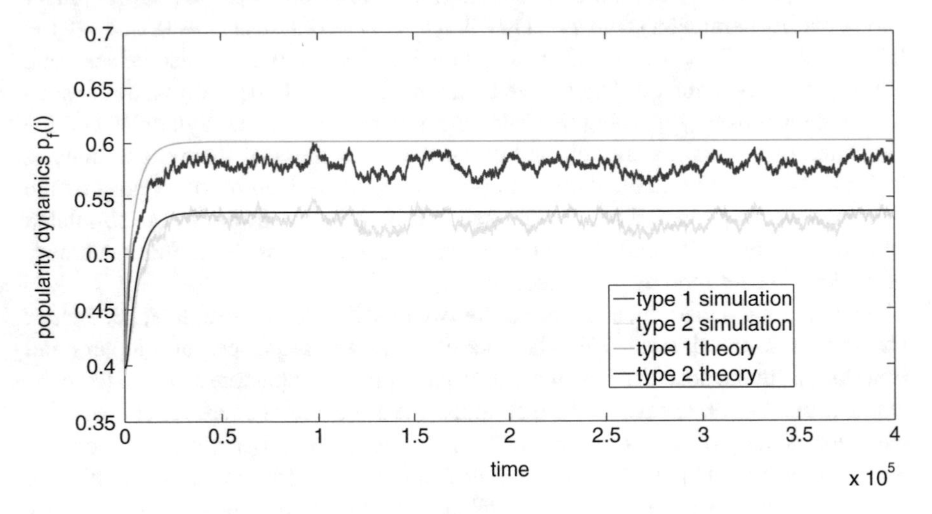

Fig. 2.9 Simulation results of the evolution dynamics for the unknown user type model with $u_{ff}(1) = 0.5$, $u_{fn}(1) = 0.8$, $u_{nn}(1) = 0.1$, $u_{ff}(2) = 0.1$, $u_{fn}(2) = 0.5$, $u_{nn}(2) = 0.3$

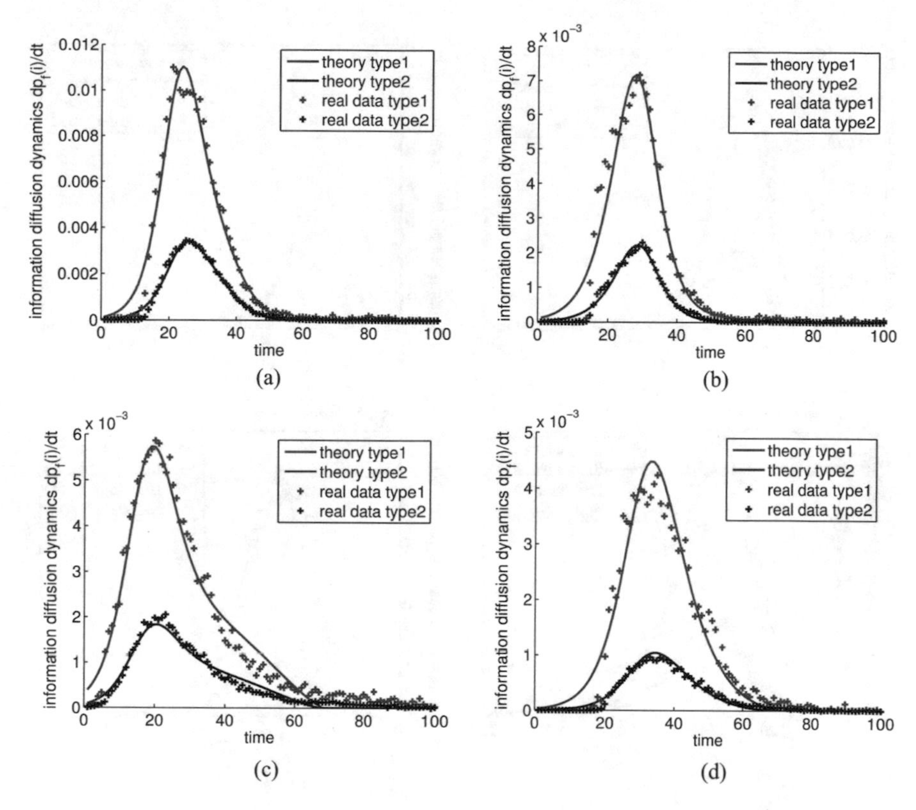

Fig. 2.10 Fitting results of the known user type model for the four popular Twitter hashtags

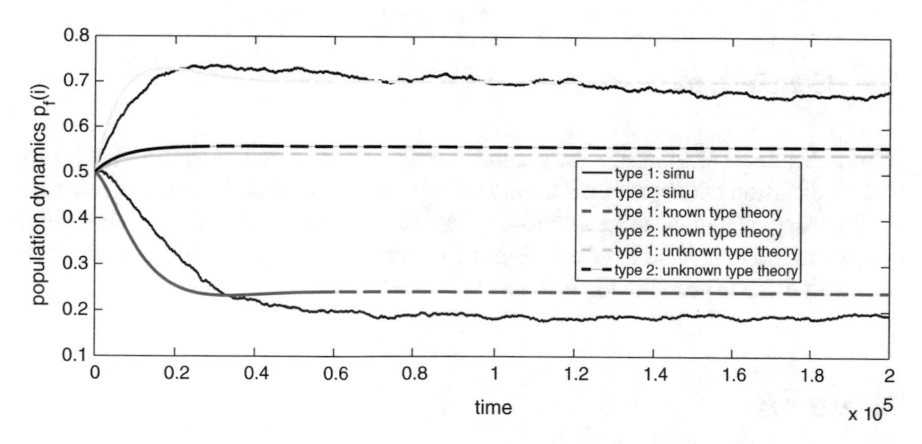

Fig. 2.11 Simulation results of the evolution dynamics for the known user type model

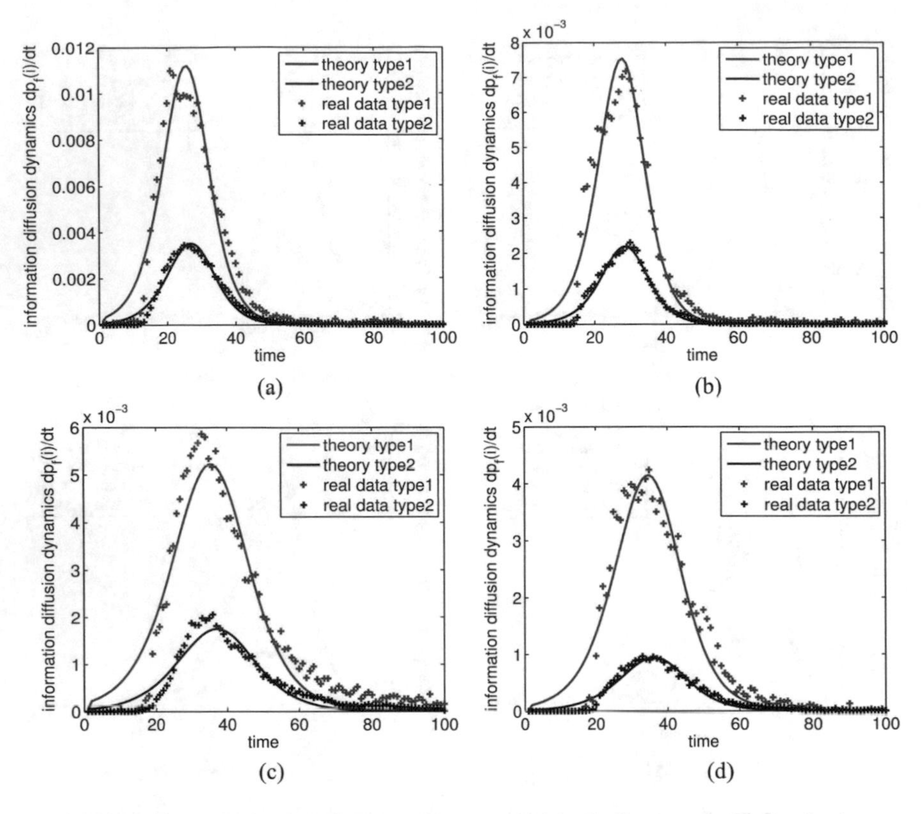

Fig. 2.12 Fitting results for the unknown user type model for the four popular Twitter hashtags

2.6 Conclusions

In this chapter, how to utilize the graphical evolutionary game to analyze the information diffusion among rational users is reviewed and discussed, with emphasis on evolutionary dynamics and evolutionary stable state. Simulation results show that the theoretical analysis of information diffusion based on graphical EGT matches well with those from synthetic data and real-world data.

References

1. C. Smith, By the numbers: 250 amazing twitter statistics. [Online]. Available: http://expandedramblings.com/index.php/march-2013-by-the-numbers-a-few-amazing-twitter-stats/ (2014)
2. D. Noyes, The top 20 valuable facebook statistics—updated october 2014. [Online]. Available: https://zephoria.com/social-media/top-15-valuable-facebook-statistics/ (2014)

3. M. Kimura, K. Saito, R. Nakano, Extracting influential nodes for information diffusion on a social network, in *Proceedings of AAAI Conference on Artificial Intelligence* (2007), pp. 1371–1376
4. S. Usui, F. Toriumi, T. Hirayama, K. Mase, Analysis of influential features for information diffusion, in *2013 International Conference on Social Computing* (2013), pp. 905–908
5. W. Yu, C. Gao, S. Guojie, X. Kunqing, Community-based greedy algorithm for mining top-k influential nodes in mobile social networks, in *Proceedings of the 16th ACM SIGKDD International Conference on Knowledge Discovery and Data Mining*, ser. KDD '10 (ACM, New York, NY, USA, 2010), pp. 1039–1048. [Online]. Available: http://doi.acm.org/10.1145/1835804.1835935
6. R. Narayanam, Y. Narahari, A shapley value-based approach to discover influential nodes in social networks. IEEE Trans. Autom. Sci. Eng. **8**(1), 130–147 (2011)
7. M. Shahsavari, A.H. Golpayegani, Finding k-most influential users in social networks for information diffusion based on network structure and different user behavioral patterns, in *2017 IEEE 14th International Conference on e-Business Engineering (ICEBE)* (2017), pp. 220–225
8. C. Damon, The spread of behavior in an online social network experiment. Science **329**(5996), 1194–1197 (2010). [Online]. Available: https://science.sciencemag.org/content/329/5996/1194
9. E. Bakshy, I. Rosenn, C. Marlow, L. Adamic, The role of social networks in information diffusion, in *Proceedings of 21st International Conference on World Wide Web (WWW)* (2012), pp. 519–528
10. Y. Jiang, C. Scott, Predicting the speed, scale, and range of information diffusion in twitter, in *Conference: Proceedings of the Fourth International Conference on Weblogs and Social Media* (2010). [Online]. Available: https://www.microsoft.com/en-us/research/publication/predicting-speed-scale-range-information-diffusion-twitter/
11. J. Yang, J. Leskovec, Modeling information diffusion in implicit networks, in *2010 IEEE International Conference on Data Mining* (2010), pp. 599–608
12. M. Ostilli, E. Yoneki, I.X. Leung, J.F. Mendes, P. Lio, J. Crowcroft, Statistical mechanics of rumor spreading in network communities, in *Proceedings of International Conference on Computer Science* (2010), pp. 2331–2339
13. C. Peng, K. Xu, F. Wang, H. Wang, Predicting information diffusion initiated from multiple sources in online social networks, in *2013 Sixth International Symposium on Computational Intelligence and Design*, vol. 2 (2013), pp. 96–99
14. C. Yuan, D. Ji, Stochastic asymptotically stability of an information diffusion model with random perturbation in social network, in *2019 IEEE 8th Joint International Information Technology and Artificial Intelligence Conference (ITAIC)* (2019), pp. 1916–1920
15. K. Masahiro, S. Kazumi, M. Hiroshi, Blocking links to minimize contamination spread in a social network. ACM Trans. Knowl. Discov. Data **3**(2), 9:1–9:23 (2009). [Online]. Available: http://doi.acm.org/10.1145/1514888.1514892
16. M.U. Ilyas, M.Z. Shafiq, A.X. Liu, H. Radha, A distributed and privacy preserving algorithm for identifying information hubs in social networks, in *2011 Proceedings IEEE INFOCOM* (2011), pp. 561–565
17. H. Kim, E. Yoneki, Influential neighbours selection for information diffusion in online social networks, in *2012 21st International Conference on Computer Communications and Networks (ICCCN)* (2012), pp. 1–7
18. A. Kuhnle, M.A. Alim, X. Li, H. Zhang, M.T. Thai, Multiplex influence maximization in online social networks with heterogeneous diffusion models. IEEE Trans. Comput. Soc. Syst. **5**(2), 418–429 (2018)
19. P. Henrique, A.J.M., G.M.A., Using early view patterns to predict the popularity of youtube videos, in *Proceedings of the Sixth ACM International Conference on Web Search and Data Mining*, ser. WSDM '13 (ACM, New York, NY, USA, 2013), pp. 365–374. [Online]. Available: http://doi.acm.org/10.1145/2433396.2433443

20. L. Weng, F. Menczer, Y.-Y. Ahn, Predicting successful memes using network and community structure, in *Proceedings of 8th International AAAI Conference on Weblogs and Social Media* (2014), pp. 535–544
21. M.G. Rodriguez, J. Leskovec, D. Balduzzi, B. Sch?lkopf, Uncovering the structure and temporal dynamics of information propagation. Netw. Sci. **2**(1), 26–C65 (2014)
22. M. Rodriguez, J. Leskovec, B. Scholkopf, Modeling information propagation with survival theory, in *Proceedings of International Conference on Machine Learning* (2013), pp. 666–674
23. M. Gomez Rodriguez, J. Leskovec, B. Schölkopf, Structure and dynamics of information pathways in online media, in *Proceedings of the Sixth ACM International Conference on Web Search and Data Mining*, ser. WSDM '13 (ACM, New York, NY, USA, 2013), pp. 23–32. [Online]. Available: http://doi.acm.org/10.1145/2433396.2433402
24. O. Yagan, D. Qian, J. Zhang, D. Cochran, Conjoining speeds up information diffusion in overlaying social-physical networks. IEEE J. Sel. Areas Commun. **31**(6), 1038–1048 (2013)
25. S. Morris, Contagion. Rev. Econ. Stud. **67**(1), 57–78 (2000)
26. C. Jiang, Y. Chen, K.J.R. Liu, Graphical evolutionary game for information diffusion over social networks. IEEE J. Sel. Top. Signal Process. **8**(4), 524–536 (2014)
27. C. Jiang, Y. Chen, K.J.R. Liu, Evolutionary dynamics of information diffusion over social networks. IEEE Trans. Signal Process. **62**(17), 4573–4586 (2014)
28. X. Cao, Y. Chen, C. Jiang, K.J. Ray Liu, Evolutionary information diffusion over heterogeneous social networks. IEEE Trans. Signal Inf. Process. Over Netw. **2**(4), 595–610 (2016)
29. H. Ohtsuki, M.A. Nowak, J.M. Pacheco, Breaking the symmetry between interaction and replacement in evolutionary dynamics on graphs. Phys. Rev. Lett. **98**(10), 108106 (2007)
30. H. Ohtsukia, M.A. Nowak, The replicator equation on graphs. J. Theor. Biol. **243**, 86–97 (2006)
31. J.W. Weibull, *Evolutionary Game Theory*, vol. 265 (The MIT Press, 1997)
32. M. Newman, Ego-centered networks and the ripple effect. Soc. Netw. **25**(1), 83–95 (2003)
33. J. Leskovec, Stanford large network dataset collection. [Online]. Available: http://snap.stanford.edu/data
34. J. Leskovec, L. Backstrom, J. Kleinberg, Meme-tracking and the dynamics of the news cycle, in *Proceedings of the 15th ACM SIGKDD International Conference on Knowledge Discovery and Data Mining*, ser. KDD '09 (ACM, New York, NY, USA, 2009), pp. 497–506. [Online]. Available: http://doi.acm.org/10.1145/1557019.1557077
35. L. Weng, F. Menczer, Y.-Y. Ahn, Predicting successful memes using network and community structure, in *8th AAAI International Conference on Weblogs and social media (ICWSM)* (2014)

Chapter 3
"Irrational" Behavior Analysis

Abstract Social networks play an important role in our daily life, and we utilize them to contact with others as well as spreading different kinds of information every day. While enjoying the convenience of social networks, we have to acknowledge that they create some security problems. Wrong, misleading or even harmful information, virus for example, is released and disseminated by malicious users over social networks, which lead to bad influences and severe consequences. Therefore, it is necessary to understand the process of information diffusion and figure out the hazard of malicious users to the whole social network. In this chapter, we employ graphical evolutionary game theory (EGT) to investigate the negative impacts caused by malicious users in information diffusion over social networks, by theoretically analyzing the population dynamics and evolutionary stable strategies. Experiments on synthetic networks, Facebook networks and real-world microblog data set are conducted, and results validate the theoretic derivation.

Keywords Information diffusion · Malicious nodes · Social networks · Security · Evolutionary game theory

3.1 Introduction

With rapid advancements of Internet, it has been an indispensable part of our life. Based on the annual report published by International Telecommunication Union (ITU), by the end of 2018 there were 51.2% of individuals (3.9 billions of people) over the world using the Internet, which means a significant achievement towards a more inclusive information society. According to the statistics in [1, 2], 30 trillion bytes of data are created every second, including approximately 6,000 tweets, more than 40,000 Google queries and more than 2 million emails, and the amount of data created is bound to increase with the development of the Internet of Things (IoT), social networks and many other new technologies.

Large quantities of information diffused over the Internet has made the world more interconnected, made our life more convenient, and also promoted the development of society. However, it creates the security challenge as well. When the information

Y. Chen and H. V. Zhao, *Behavior and Evolutionary Dynamics in Crowd Networks*, Lecture Notes in Social Networks, https://doi.org/10.1007/978-981-15-7160-2_3

is wrong or misleading, people might be deceived by its content and then become the spreaders, leading to a vicious circle. Moreover, if it is harmful information, virus for example, it would cause severe consequences and incalculable damages. For instance, in 2000 the virus "ILOVEYOU" outbroke worldwide, starting from an email message with the subject line "ILOVEYOU" and an attachment to damage the local machine and copy this email to more people. It was reported that such an event caused the loss of more than 2 billion dollars [3, 4]. Not only spread by emails, now social applications such as Facebook, Instagram, Twitter and Wechat have become prevalent platforms for malicious users to spread rumours or even malicious information. Therefore, it is of crucial importance to model information diffusion process over social networks with malicious nodes, and quantify their negative impacts to the whole network.

There are many works on modeling the information diffusion, which could be generally classified into two categories. The first category explores from the macroscopic aspect, usually adopting machine learning or data mining techniques to predict the dynamics or properties of network. Based on historical information, [6–12] investigated current and future characterization of the dynamics of information diffusion on social media, such as YouTube [5, 6], Twitter [7, 8] and blog [9–11]. A matrix factorization based predictive model was proposed and the gradient descent algorithm was utilized to optimize objective function in [12], while in [13] authors proposed a model based on a physical radiation energy transfer mechanism to predict the diffusion graph of a certain contagion. Tsai et al. in [14] used a rank-learning based data-driven approach to study the diffusion of preference on social networks. A K-center method was proposed in [15] to realize multi-source identification of information diffusion and the corresponding infection regions in general networks. In [16], the authors compared different heuristic influence maximization techniques and proposed a machine learning based approach to find the spread of information in the network. Focusing on macro-scale diffusion structures, Yang et al. aimed to reveal the correlation between public opinion and diffusion structures by combining vocabulary knowledge with supervised machine learning algorithms [17].

The second category focuses on micro exploration. In other words, it pays more attention to the decisions and motivation of individuals. The problem of predicting dynamic trends with each user's activeness under dynamic activeness model was studied in [18]. Lee et al. in [19] proposed a probabilistic model to estimate the probability of a user's decision with the naive Bayes classifier. Based on game theory, authors proposed a framework to study the competition between firms who aim to maximize consumers' adoption in social networks in [20, 21]. ProfileRank was proposed in [22], which measured content relevance and user influence based on random walks over a user-content bipartite graph. Authors in [7, 23–25] modeled information diffusion by defining different objective functions for each user and then solving the corresponding minimization or maximization problem.

Evolutionary game theory (EGT) could be a satisfactory tool to study the information propagation and understand details of users' interactions. In [26], the authors studied the adaptive network from the game theoretic perspective, and it was shown that the proposed framework could provide a general framework to unify the existing distributive adaptive network algorithms. An evolutionary dynamics model was proposed in [27], where users determined actions among multiple behaviors. In [28–30], an evolutionary game-theoretic framework was proposed to model the dynamic information diffusion process in social networks, where the authors in [28, 29] focused on the final stable state, while in [30] the emphasis was on the evolutionary dynamics. Cao et al. in [31] extended this EGT model to heterogeneous social networks where there are different types of nodes. While these works modeled the process of information diffusion well, most of existing solutions do not take irrational users in networks into account, who generally don't follow particular principles to update their strategies.

Considering negative aspects in information diffusion, there are also some works related to misinformation or rumour propagation. Authors in [32] introduced a multi-agent information diffusion model based on Susceptible-Infected-Recovered (SIR) to study the false rumor diffusion, while in [33] an information propagation model was developed to differentiate rumors through the distinct propagation patterns. Zhao and Chen in [34] analyzed the propagation regular of derivative rumor by constructing a linear threshold model. In [35, 36], authors proposed a subjective logic based opinion model and variant of the SIR epidemic model, where multiple pieces of related information were considered to model the evolution rule and agents were assumed to share opinions. In this chapter, we mainly focus on the propagation of one piece of information and assume that users only share the strategies, which is more suitable for large-scale information propagation as users may not realize others' beliefs and uncertainty. Most of existing works lay emphasis on the false or negative information itself rather than malicious behaviors of nodes. Nevertheless, malicious nodes, whose actions are quite different from rational nodes, commonly exist in social networks. Their unfavoured motivations, such as attacking the social networks and/or spreading the detrimental information to enlarge their influences over network, may lead to severe security problem and dramatic economic loss.

In this chapter, based on graphical EGT, we estimate the hazard impact of malicious nodes in information diffusion over social networks. Specifically, we first employ graphical EGT to formulate the process. Due to the existence of malicious nodes, rational nodes could be divided into two types: type I nodes which are directly connected to the malicious nodes and type II nodes which are not directly connected to the malicious nodes. Then, we theoretically analyze the evolution dynamics and the corresponding evolutionary stable strategies (ESSs) of type I nodes and type II nodes, respectively. Finally, we conduct simulations on synthetic networks and Facebook networks as well as experiments on real-world microblog data set to validate the theoretic derivations, and results show that the proposed method fits well with actual conditions when there are malicious users in the social network.

3.2 Evolutionary Game Theory

Starting from the problem of explaining animals' altruistic behaviour in contests for resources, evolution has converted from a biological concept into a useful tool applied in different research areas such as economics, finance and securities. Developed from classical game theory, EGT doesn't require users to act fully rationally, and it doesn't demand that global complete details are available as well. The key insight of EGT is the interactions of multiple players when deciding their behaviors, and the success of a certain behavior depends on how the corresponding players interact with others. Accordingly, the fitness of an individual cannot be measured in isolation but evaluated in view of surrounding population.

Without loss of generality, a simple game with two types of players–type I and type II is considered, and each type has a strategy set $i \in S(I)$ and $j \in S(II)$, respectively. The payoffs for a type I player with strategy $i \in S(I)$ and a type II player with strategy $j \in S(II)$ are denoted as a_{ij} and b_{ij} when they are interacting. So payoffs in all cases could be represented as $m \times n$-matrices $A = [a_{ij}]$ and $B = [b_{ij}]$, in which n and m are the cardinalities of the sets of strategies [37]. In graphical EGT [38, 39], the fitness of players determines their strategy updating rule, and now we give its definition as follows

$$\pi = (1 - \alpha) + \alpha U \tag{3.1}$$

where $\alpha \in (0, 1)$ is the selection strength and U is the payoff (a_{ij} or b_{ij}). π could be regarded as the linear combination of payoff U and the baseline fitness, which is always set to be 1. In the literature [29–31], α is a very small value, corresponding to weak selection. Hence, we could denote the fitness for type I players and type II players as $\pi_I = (1 - \alpha) + \alpha a_{ij}$ and $\pi_{II} = (1 - \alpha) + \alpha b_{ij}$, respectively.

Based on fitness, players follow a specific rule to update their strategies, including adopting a new strategy and remaining current strategy. As time passes, the game would trend towards equilibrium and players would choose to adopt evolutionary stable strategy (ESS) gradually [40]. When ESSs are adopted by the whole population, then a small group of "invaders" with any alternate strategies are doomed to have lower fitness than other users, and eventually die off with a high probability. Mathematically, ESS could be described as: a strategy $\mathbf{p}^* \in \mathbf{S}(I)$ is evolutionarily stable for type I players when for every $\mathbf{p} \in \mathbf{S}(I)$ with $\mathbf{p} \neq \mathbf{p}^*$, the following conditions hold [37]

$$\begin{aligned} &(a)\ \mathbf{p}^T A\mathbf{p}^* \leq \mathbf{p}^{*T} A\mathbf{p}^*, \\ &(b)\ If\ \mathbf{p}^T A\mathbf{p}^* = \mathbf{p}^{*T} A\mathbf{p}^*,\ then\ \mathbf{p}^T A\mathbf{p} < \mathbf{p}^{*T} A\mathbf{p}. \end{aligned} \tag{3.2}$$

The first condition means that $\mathbf{p}^*$ is a Nash equilibrium, i.e., residents' interactions are always better than invaders' interactions, while the second condition signifies that if the invader does as well as the resident against the resident, then it does worse than the resident against the invader.

To find the ESS, one universal way is by means of evolutionary dynamics. Let $\mathbf{x} = [x_1, x_2, \ldots, x_i, \ldots, x_m]$ be the system states, where each element x_i is the proportion of players adopting strategy i. Then we could derive the ESS as

$$\mathbf{x}^* = \arg\left(\frac{d\mathbf{x}}{dt} = 0\right). \tag{3.3}$$

3.2.1 Strategy Updating Rules

According to [41], there are three typical strategy updating rules: birth-death (BD), death-birth (DB) and imitation (IM).

1. BD updating rule: a player is chosen for reproduction with the probability being proportional to fitness (Birth process). Then, the chosen player's strategy replaces one neighbor's strategy with uniform probability (Death process).
2. DB updating rule: a random player is chosen to abandon his/her current strategy (Death process). Then, the chosen player adopts one of his/her neighbors' strategies with the probability being proportional to their fitness (Birth process).
3. IM updating rule: each player either adopts the strategy of one neighbor or remains his/her current strategy, with the probability being proportional to fitness.

It has been proved that the analyses and results of these three updating rules are similar in [30]. Therefore, without loss of generality, in this chapter we only analyze the condition of DB updating rule, as shown in Fig. 3.1.

3.2.2 Graphical Evolutionary Game Formulation

Figure 3.2 shows the structure of social network in our model. Represented as an undirected graph, this social network consists of nodes denoting users and edges denoting their mutual connections. Among all nodes, there are $M + N$ that are rational and update their strategies according to the predefined DB updating rule, and f_{max}

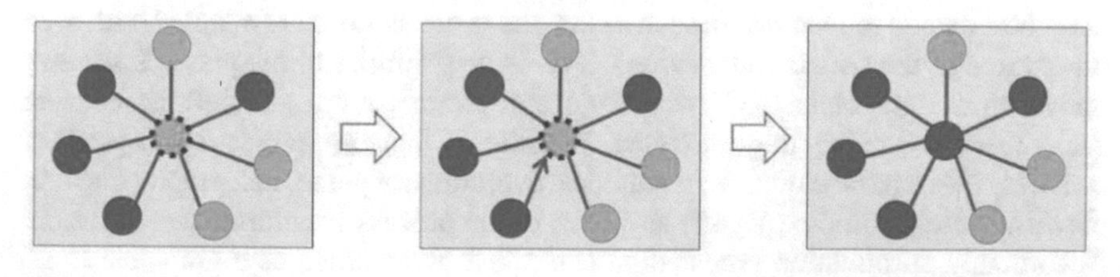

Fig. 3.1 An illustration of the DB updating rule: the node with red dashed line is the one chosen to abandon its current strategy, blue arrow represents the strategy changing process, blue lines are mutual correlations of players, and different colors inside nodes represent different strategies

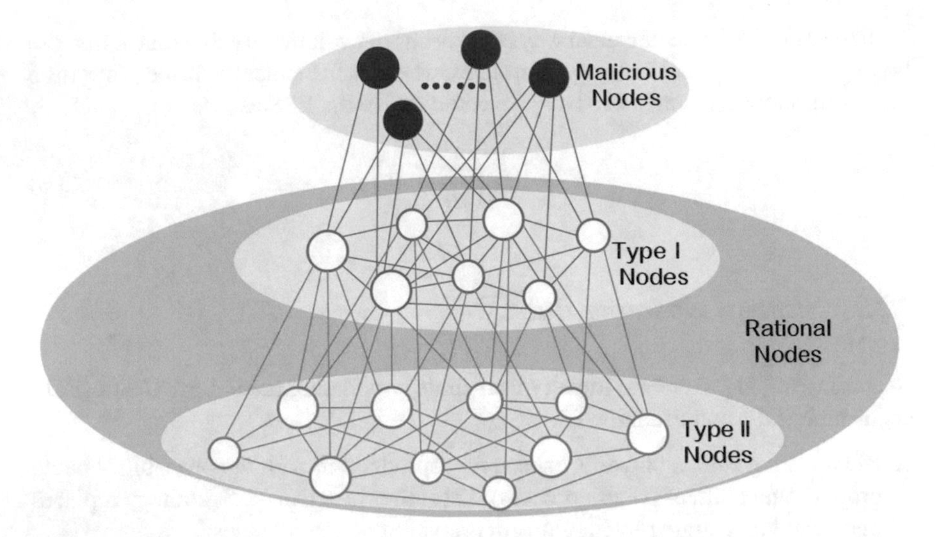

Fig. 3.2 An illustration of the social network with type I nodes, type II nodes, and malicious nodes

that are malicious and use a fixed malicious strategy. The target of rational nodes is to maximize their fitness, while malicious nodes don't have specific goals and they arc often paid to spread harmful information. Based on connections to malicious nodes, rational nodes could be further divided into two types: M type I nodes which are directly linked to malicious nodes, and N type II nodes which are not directly linked. For each rational node, there are k rational nodes as their neighbor nodes. k obeys the distribution $\lambda(k)$, which means that when randomly choosing one rational node, the probability of the chosen node with k rational neighbors is $\lambda(k)$. While for each type I rational node, it has extra f malicious neighbor nodes apart from k rational neighbors with probability distribution $\mu(f)$. As a result, every type I rational node has $(k + f)$ neighbors while type II rational node only has k neighbors.

When a user or a small group releases a new piece of information, the propagation dynamics and final stable state of the information are largely up to other users, i.e., their decisions on whether to forward. Several factors determine whether the user reposts the information, such as his/her interests and neighbors' reactions since the center node would forward information with high probability if most of his/her neighbors decide to forward. In such a case, the users' actions are coupled with each other through their social interactions. This is very similar to the player's strategy update in the graphical EGT, where players' strategies are also influenced with each other through the graph structure. In addition, there are perfect corresponding contents of five basic elements in graphical evolutionary game, i.e., graph structure, players, strategy, fitness (payoff) and ESS, in the process of information diffusion. Accordingly, information propagation could just be regarded as a graphical EGT. Based on the correspondence, in this paper, the set of users' strategy is defined as $S = \{S_f, S_n\}$, where forwarding information is denoted as S_f and not forwarding is denoted as S_n. For a rational user, he/she chooses the strategy based on the DB updating rule, while for a malicious user, we assume that he/she could only adopt S_f, mimicking the scenarios that some people are employed to spread misleading information or hackers deliberately spread computer virus. In general, when we first contact with some new information, the reliability of the disseminators aren't able to be distinguished initially and thus they are always treated equally. In such a case, the payoff matrix is the same for every type of rational users because the information on whether the neighbor is malicious or which type the neighbor belongs to is not accessible. Since there is only one payoff matrix for all rational users and strategy set only has two elements, the payoff matrix can be written as a 2×2 matrix:

$$\begin{array}{cc} & \begin{array}{cc} S_f & S_n \end{array} \\ \begin{array}{c} S_f \\ S_n \end{array} & \begin{pmatrix} u_{ff} & u_{fn} \\ u_{nf} & u_{nn} \end{pmatrix} \end{array} \tag{3.4}$$

where u_{ff}, u_{fn}, u_{nn} denotes the payoffs for two nodes when they both adopt S_f, one adopts S_f while the other adopts S_n, and they both adopt S_n, respectively. Apparently, $u_{fn} = u_{nf}$, i.e., the payoff matrix is symmetric.

To evaluate the hazard of malicious nodes on the whole network, we use population state p_f to indicate the current condition of information diffusion, which is defined as the proportion of rational users adopting S_f among all rational users. In the same way, we define population state for each type as p_{f1} and p_{f2}. The difference of population state between two adjacent time indexes is population dynamics, denoted as $\dot{p_f}$, $\dot{p_{f1}}$, $\dot{p_{f2}}$, respectively. Due to the existence of malicious users, we analyze type I nodes and type II nodes respectively in the following.

3.3 Theoretic Analysis

In this section, we analyze the evolutionary dynamics and corresponding ESSs of users of two types to discern malicious users' effects. The procedures of theoretic analysis are shown in Fig. 3.3.

3.3.1 Evolutionary Dynamics of Type I Nodes

Based on the basic form (3.1), the fitness for the type I node could be denoted as

$$\pi_{f1} = 1 - \alpha + \alpha \left[k_f u_{ff} + (k + f - k_f) u_{fn}\right], \tag{3.5}$$

and

$$\pi_{n1} = 1 - \alpha + \alpha \left[k_f u_{fn} + (k + f - k_f) u_{nn}\right], \tag{3.6}$$

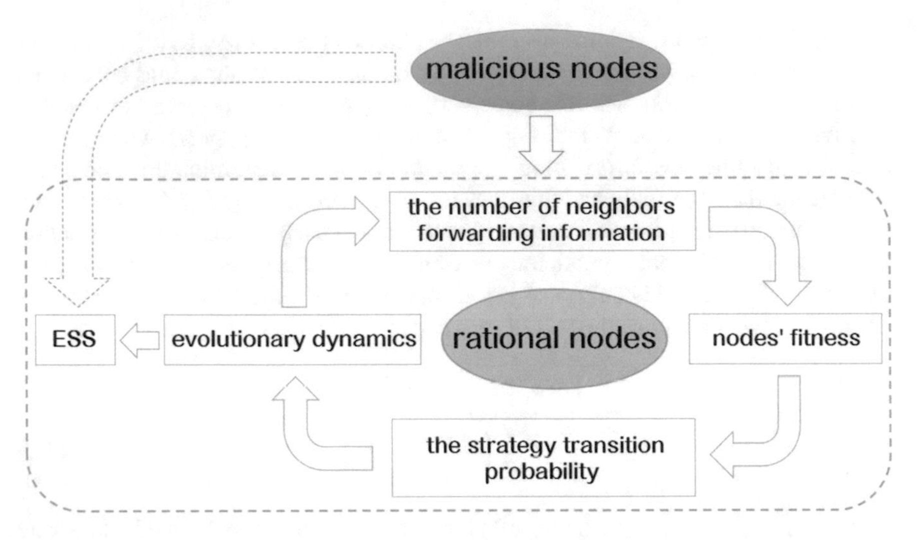

Fig. 3.3 System model: an illustration of how malicious nodes influence the evolution dynamics and ESS

where k_f is the number of nodes with strategy S_f (including malicious nodes) in all neighbors, π_{f1} and π_{n1} are the fitness of nodes adopting S_f and S_n, respectively.

During strategy updating, users may deviate from his/her current strategy with probability proportional to the fitness of that strategy. The probabilities transited to strategy S_f and strategy S_n can be derived respectively as

$$P_{to\_f1} = \frac{k_f \pi_{f1}}{k_f \pi_{f1} + (k + f - k_f)\pi_{n1}}, \tag{3.7}$$

and

$$P_{to\_n1} = \frac{(k + f - k_f)\pi_{n1}}{k_f \pi_{f1} + (k + f - k_f)\pi_{n1}}. \tag{3.8}$$

According to DB updating rule, one of $(M + N)$ nodes would be randomly selected to update in each round. The scenario in (3.7) only happens when the chosen node is a type I node with S_n, and the probability is $(1 - p_{f1})M/(M + N)$. Similarly the scenario in (3.8) happens with probability $p_{f1}M/(M + N)$ when the chosen node is a type I node with S_f. This node may deviate from S_f to S_n, bringing about the proportion of type I users adopting S_f, also known as p_{f1}, decreasing by $1/M$ with probability

$$Prob\left(\Delta p_{f1} = -\frac{1}{M}\right) = \frac{M}{M + N} p_{f1} P_{to\_n1}. \tag{3.9}$$

On the contrary, when the selected one is a type I node adopting S_n, p_{f1} may increase by $1/M$ with probability

$$Prob\left(\Delta p_{f1} = \frac{1}{M}\right) = \frac{M}{M+N}(1 - p_{f1})P_{to\_f1}. \tag{3.10}$$

Then combining two scenarios above, the evolutionary dynamics of type I nodes can be derived as follows

$$\begin{aligned}\dot{p}_{f1} &= \mathbf{E}\left[Prob\left(\Delta p_{f1} = -\frac{1}{M}\right)\left(-\frac{1}{M}\right) + Prob\left(\Delta p_{f1} = \frac{1}{M}\right)\frac{1}{M}\right] \\ &= \frac{1}{M+N}\mathbf{E}\left[\frac{k_f}{k+f} - p_{f1} + \alpha \cdot \frac{-\Phi k_f^3 + (\Phi - \Phi_n)(k+f)k_f^2 + (k+f)^2\Phi_n k_f}{(k+f)^2}\right],\end{aligned} \tag{3.11}$$

where $\Phi = u_{ff} - 2u_{fn} + u_{nn}$ and $\Phi_n = u_{fn} - u_{nn}$.

In (3.11), Maclaurin series $\frac{a+b\alpha}{c+d\alpha} = \frac{a}{c} + \frac{bc-ad}{c^2}\alpha + O(\alpha)$ is used to transfer fraction to polynomial, which facilitates the computation of expectation. The higher order term $O(\alpha)$ in (3.11) is omitted because α is a very small value with assumption of weak selection.

For the center node, their neighbors' strategies are not correlative, so the rational neighbors' strategies could be modeled as a Bernoulli sequence. In this way, k_f and p_f are connected and thus we can fully understand the dynamics of type I nodes. Then we consider the probability of encountering a rational neighbor adopting S_f as $p_{total\_f} = \frac{M_f+N_f+f_{max}}{M+N+f_{max}} \approx p_f + \frac{f_{max}}{M+N+f_{max}}$ rather than p_f in [31], since malicious nodes affect all rational nodes in the network and they should be taken into account in the calculation of the proportion. M_f and N_f denote the number of nodes with strategy S_f among type I nodes and type II nodes, respectively. The approximation is reasonable under the assumption that f_{max} is quite small compared with $M + N$, i.e., there is only a small amount of malicious nodes among whole network. When the center node has k rational neighbors, there are $k_f - f$ nodes with S_f and $(k + f - k_f)$ nodes with S_n, and the probability of such a configuration is

$$\theta(k, k_f - f) = \binom{k}{k_f - f} p_{total\_f}^{k_f - f}\left(1 - p_{total\_f}\right)^{k+f-k_f}. \tag{3.12}$$

With (3.12), considering extra f malicious neighbors, the moments of k_f can be derived as follows

$$\begin{aligned}\mathbf{E}(k_f) &= kp_{total\_f} + f, \\ \mathbf{E}(k_f^2) &= (k^2 - k)p_{total\_f}^2 + (2fk + k)p_{total\_f} + f^2, \\ \mathbf{E}(k_f^3) &= k(k-1)(k-2)p_{total\_f}^3 + 3(f+1)k(k-1)p_{total\_f}^2 \\ &\quad + (3f^2 + 3f + 1)kp_{total\_f} + f^3.\end{aligned} \tag{3.13}$$

Combining (3.11) with (3.13), we could obtain the evolution dynamics of type I nodes $\dot{p}_{f1}$ as

$$\dot{p}_{f1} = \sum_{k}\sum_{f=1}^{f_{max}} \mu(f)\lambda(k)\left\{\frac{f(1-p_{f1})+k(p_{total\_f}-p_{f1})}{(M+N)(k+f)} + \frac{\alpha k(1-p_{total\_f})}{(M+N)(k+f)^2}\right.$$
$$\Big[(k-1)(k-2)\Phi p_{total\_f}^2 + [(2f+1)(k-1)\Phi + (k+f)(k-1)\Phi_n]p_{total\_f}$$
$$\left. + f^2\Phi + f(k+f)\Phi_n\Big]\right\}. \tag{3.14}$$

From (3.14), we observe that both p_{f1} and p_f determine $\dot{p}_{f1}$, which means that nodes are affected not only by those of same type, but also other nodes in the network.

3.3.2 Evolutionary Dynamics of Type II Nodes

Compared with the situation of type I nodes stated before, there aren't direct-connected malicious neighbors in the situation for type II nodes, which means that the number of neighbors of type II nodes is k rather than $(k+f)$. In such a case, the fitness could be written as

$$\pi_{f2} = 1-\alpha+\alpha\left[k_f u_{ff} + (k-k_f)u_{fn}\right], \tag{3.15}$$

and

$$\pi_{n2} = 1-\alpha+\alpha\left[k_f u_{fn} + (k-k_f)u_{nn}\right]. \tag{3.16}$$

In the same way, the probabilities of strategy transition P_{to_f2} and P_{to_n2} can be denoted as

$$P_{to\_f2} = \frac{k_f\pi_{f2}}{k_f\pi_{f2} + (k-k_f)\pi_{n2}}, \tag{3.17}$$

and

$$P_{to\_n2} = \frac{(k-k_f)\pi_{n2}}{k_f\pi_{f2} + (k-k_f)\pi_{n2}}. \tag{3.18}$$

Then the evolutionary dynamics for type II nodes, i.e., $\dot{p}_{f2}$, can be derived according to [31] as

$$\dot{p}_{f2} = \sum_{k}\lambda(k)\cdot$$
$$\left\{\frac{p_{total\_f}-p_{f2}}{M+N} + \frac{\alpha(k-1)p_{total\_f}(1-p_{total\_f})}{(M+N)k}\left[(k-2)\Phi p_{total\_f} + \Phi + k\Phi_n\right]\right\}. \tag{3.19}$$

It could be observed that on account of malicious nodes' absence, the resulting formula is more precise, which simplifies the solution to ESS as shown in the next subsection.

3.3.3 ESS Analysis

To obtain the evolutionary dynamics of the whole network, we integrate (3.14) and (3.19) linearly in proportion to the percentage of type I nodes and II nodes among all rational nodes as

$$\begin{aligned}\dot{p}_f &= \frac{M}{M+N}\dot{p}_{f1} + \frac{N}{M+N}\dot{p}_{f2}\\ &= \sum_k \frac{\lambda(k)}{(M+N)^2}(1-p_{total\_f})(ap_{total\_f}^2 + bp_{total\_f} + c)\\ &= \sum_k \frac{\lambda(k)}{(M+N)^2}\left(1 - p_f - \frac{f_{max}}{M+N+f_{max}}\right)\\ &\times\left[ap_f^2 + \left(\frac{2af_{max}}{M+N+f_{max}} + b\right)p_f + a\left(\frac{f_{max}}{M+N+f_{max}}\right)^2 + \frac{bf_{max}}{M+N+f_{max}} + c\right],\end{aligned} \tag{3.20}$$

where

$$\begin{aligned}a &= \alpha(k-1)(k-2)\left(\sum_{f=1}^{f_{max}} \frac{kM\mu(f)\Phi}{(k+f)^2} + \frac{N\Phi}{k}\right),\\ b &= \alpha(k-1)\left[\sum_{f=1}^{f_{max}} \frac{\mu(f)kM}{k+f}\left(\frac{(2f+1)\Phi}{k+f} + \Phi_n\right) + \frac{N\Phi}{k} + N\Phi_n\right],\\ c &= \sum_{f=1}^{f_{max}} \frac{M\mu(f)}{k+f}\left(\frac{\alpha kf^2}{k+f}\Phi + \alpha kf\Phi_n + f\right).\end{aligned} \tag{3.21}$$

Notice that the a, b, c in (3.21) are coefficients of the quadratic equation in (3.20). From (3.20), we could find that by setting $\dot{p_f} = 0$, there are three possible ESSs, i.e., $p_f^* = 1 - \frac{f_{max}}{M+N+f_{max}}$ and two roots to the quadratic equation $ap_f^{*2} + \left(\frac{2af_{max}}{M+N+f_{max}} + b\right)p_f^* + a\left(\frac{f_{max}}{M+N+f_{max}}\right)^2 + \frac{bf_{max}}{M+N+f_{max}} + c = 0$, which both lie between 0 and 1.

Different from the results in [31] that when payoffs satisfy $u_{nn} > u_{fn}$ ESS would be 0, in (3.20) none of three ESSs is equal to zero. This is because the existence of malicious nodes creates an illusion that there are quite a few neighbors adopting S_f in spite of biggish u_{nn}, directly impacting on type I nodes to deviate from current strategy to S_f and then indirectly impacting on type II nodes. It indicates that with malicious nodes in social network, the proportion of users adopting S_f inclines. The more malicious nodes in the network, the larger ESSs would be, which could also be obtained from the roots of the quadric equation in (3.20). When there is no malicious node, the results would reduce back to those in [31], as $f_{max} = 0$, $c = 0$ and $\mu(0) = 1$.

3.4 Simulation and Experiment Results

In this section, we validate the proposed EGT model through simulations and experiments, and evaluate the harm of malicious users to the entire social network. We first simulate in synthetic networks with predefined parameters. Then we conduct simulations on real Facebook network and compare theoretic analysis with simulated results. Finally, retweet behaviors in a microblog data set are studied to verify the effectiveness of the method.

3.4.1 Synthetic Networks

First of all, we simulate a uniform degree network with degree $k = 25$. Assuming that malicious neighbors of type I nodes are evenly distributed, i.e., $\mu(f) = 1/f_{max}$, $1 \leq f \leq f_{max}$. We first evaluate the performance with different number of malicious nodes, and the results are shown in Fig. 3.4. For the network, we generate 1500 rational nodes, including 500 type I nodes and 1000 type II nodes, and initialize their strategies: 30% adopting S_f and 70% adopting S_n. The payoff matrix is set as $PM1: u_{ff} = 0.3, u_{fn} = 0.8, u_{nn} = 0.2$ and weak selection parameter α is 0.025. Figure 3.4a, b, c show the evolutionary population states p_{f1}, p_{f2} and p_f under the scenarios $f_{max} = 0$, $f_{max} = 5$ and $f_{max} = 10$. From these figures we could observe that the theoretic results fit well with the simulation results, about 0.75% relative error with 5 malicious users and 0.54% relative error with 10 malicious users.

From Fig. 3.4, it can also been observed that as the number of malicious nodes increases from 0 to 10, the proportion of rational nodes with S_f is increasing for both type I and type II nodes. For example, in Fig. 3.4c, compared with the baseline, i.e. the black curve, the ESS of p_f increases by 17% when there are 5 malicious users and by 28.3% when there are 10 malicious users. From Fig. 3.4a, b, c, we could see that the proportion of type I nodes adopting S_f is higher than that of type II nodes, which is because type I nodes are apt to adopt S_f under malicious users' direct influence. Affected by type I nodes, some type II nodes tend to change their current strategies to S_f. Therefore, ESSs are largely enhanced from the baseline with malicious users' direct and indirect influence. Note that the baselines in Fig. 3.4a, b, c are identical, because there is no difference between all rational nodes without malicious nodes in the network.

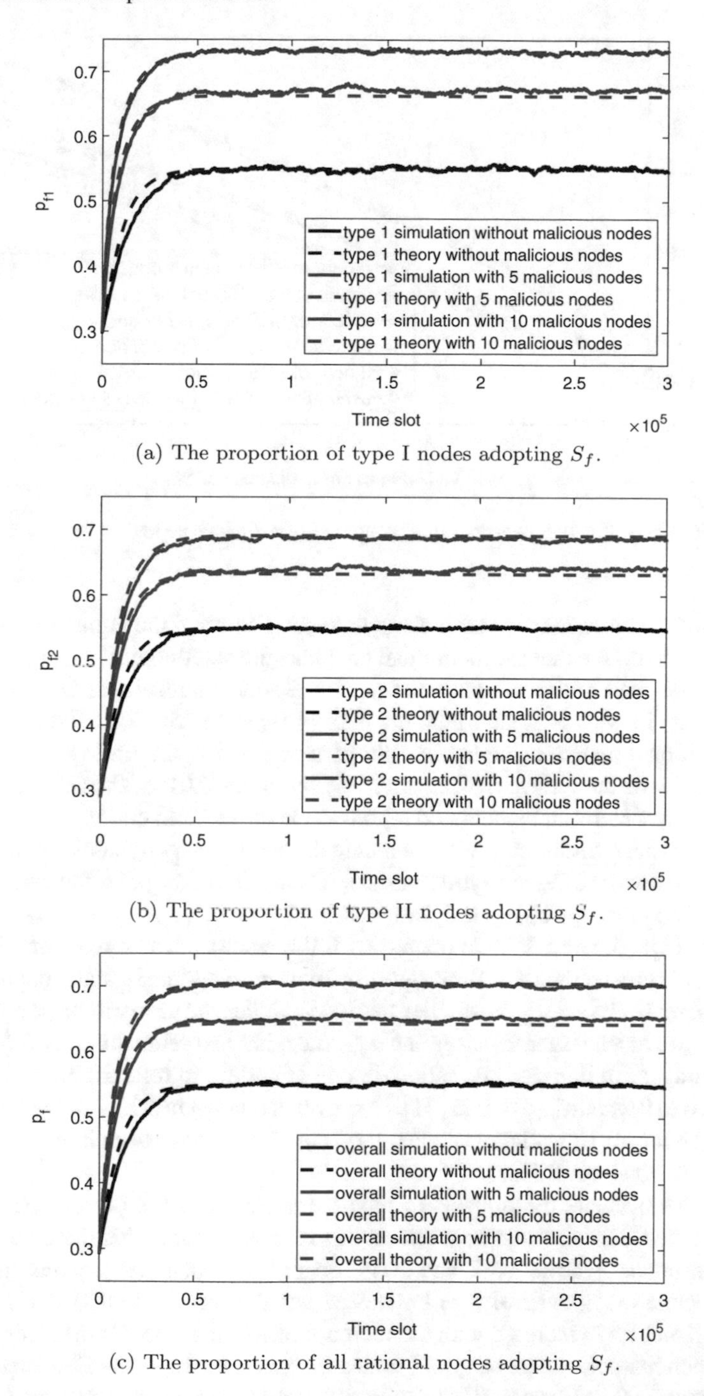

(a) The proportion of type I nodes adopting S_f.

(b) The proportion of type II nodes adopting S_f.

(c) The proportion of all rational nodes adopting S_f.

Fig. 3.4 Simulation results of the evolution dynamics under the payoff matrix PM1: $u_{ff} = 0.3$, $u_{fn} = 0.8$, $u_{nn} = 0.2$

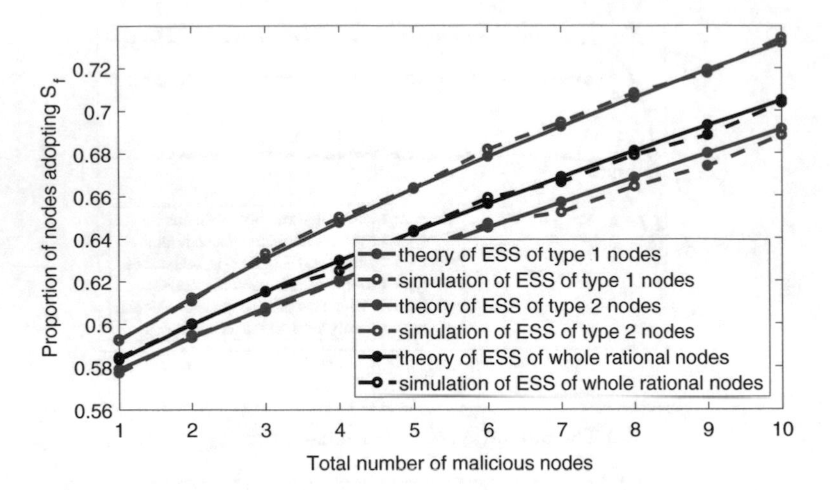

Fig. 3.5 The ESS of networks with different number of malicious nodes

Figure 3.5 illustrates the relationship between ESSs and the number of malicious nodes. We could see that the theoretical analysis, the solution of the equation (3.20), again agrees well with the simulation results. Similar to results in the Fig. 3.4, the ESS of type I nodes is greater than the ESS of type II node. It can be seen that for each additional malicious node, the ESS of type I nodes, the ESS of type II nodes, and the ESS of the entire network increase by about 0.0155, 0.0125, and 0.0135, respectively, which can be obtained by the slope of the solid curve.

Then, we perform simulations to evaluate the evolutionary dynamics with 10 malicious nodes under different payoff matrices. The results of the payoff matrices $PM2$: $u_{ff} = 0.2, u_{fn} = 0.8, u_{nn} = 0.5$ and $PM3 : u_{ff} = 0.1, u_{fn} = 0.1, u_{nn} = 0.8$ are shown in Figs. 3.6 and 3.7. Compared with the results of the payoff matrix $PM1$, where u_{ff} is greater than u_{nn}, the proportion of users adopting S_f under payoff matrix $PM2$ decreases when u_{ff} is smaller than u_{nn}. Under the scenario of payoff matrix $PM3$, u_{nn} is much larger than u_{ff} and u_{fn}, meaning that retransmitting information is discouraged. In this case, no rational node is willing to forward information and ESS tends to 0 when $f_{max} = 0$ in [31]. However, as shown in 3.7, under the direct and indirect effects of 10 malicious nodes, p_{f1}, p_{f2} and p_f have been largely increased, resulting in non-zero ESSs.

Next, we evaluate the impacts of payoff matrices and the percentage of type I nodes on the ESSs of whole network. In this case we assume that there are totally 5 malicious nodes in the network, and the number of rational nodes remains unchanged at 1500 while the number of type I nodes M varies from 200 to 800. As we can see from Fig. 3.8, ESSs increase with the increase of M for all payoff matrices. Specifically, when the percentage of type I nodes increases by 10%, the ESS increases by approximately 0.041 under $PM1$, approximately 0.034 under $PM2$, and approximately 0.025 under $PM3$. This indicates that the more connections between malicious nodes and reasonable nodes, the better the information spread.

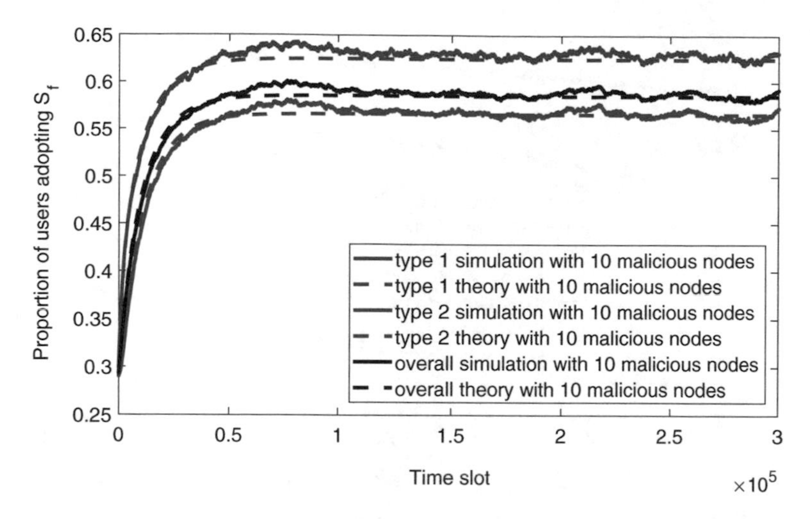

Fig. 3.6 Simulation results of the evolution dynamics under the payoff matrix PM2: $u_{ff} = 0.2$, $u_{fn} = 0.8$, $u_{nn} = 0.5$

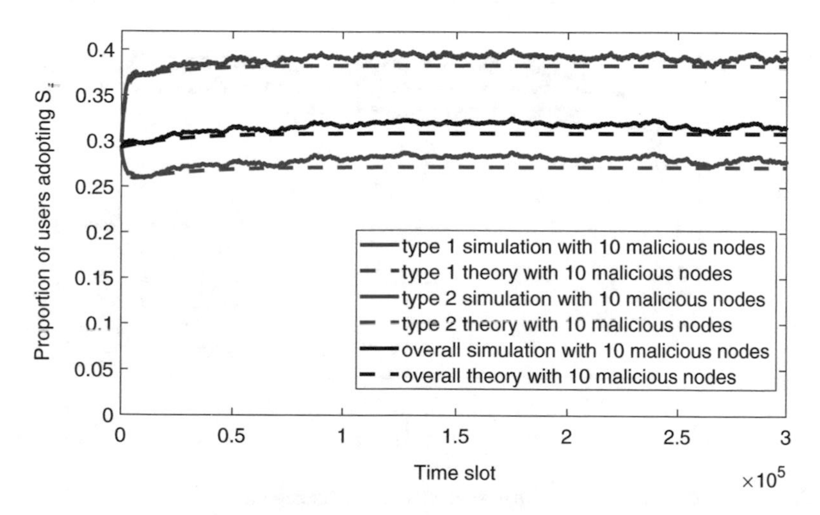

Fig. 3.7 Simulation results of the evolution dynamics under the payoff matrix PM3: $u_{ff} = 0.1$, $u_{fn} = 0.1$, $u_{nn} = 0.8$

Figure 3.9 shows the effect of weak selection parameter α on p_f, with payoff matrix set as $PM1$ and the number of malicious users set as 10. The implication of α is the relative contribution of the interaction between nodes to fitness. Therefore, a larger α indicates that in each round of strategy updating, the surrounding environment is more important for the fitness. Three different cases are considered here: $\alpha = 0.02$, $\alpha = 0.025$ and $\alpha = 0.03$, which all satisfy weak selection. As shown in Fig. 3.9, we can observe that the ESS would decrease with α increasing, which could also be obtained from (3.20) and (3.21). From the results, it could be learned in practice that

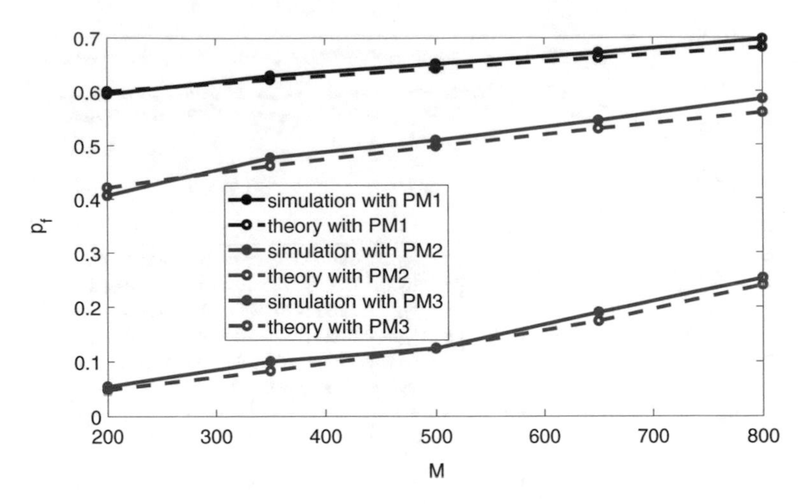

Fig. 3.8 The comparison with different M under three different payoff matrices

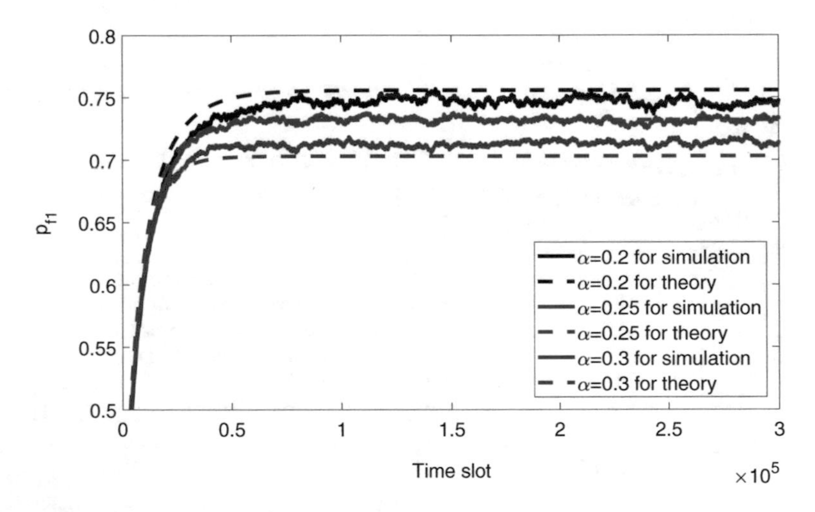

Fig. 3.9 The comparison with different weak selection parameter α

maintaining vigilance towards received information, i.e., a relatively large α, is a better choice. By this way p_f would decrease, correspondingly meaning the reduced diffusion of detrimental information or computer virus.

At last we validate the proposed model over non-uniform degree networks, including the Erdős-Rényi random network and Barabási-Albert scale-free network, as shown in Figs. 3.10 and 3.11. We assume that payoff matrix is set as $PM1$ and $f_{max} = 10$. For the Erdős-Rényi random (ER) network, the degree obeys a Poisson distribution, i.e.,

$$\lambda_{ER}(k) = \frac{e^{-\bar{k}}\bar{k}^k}{k!}, \tag{3.22}$$

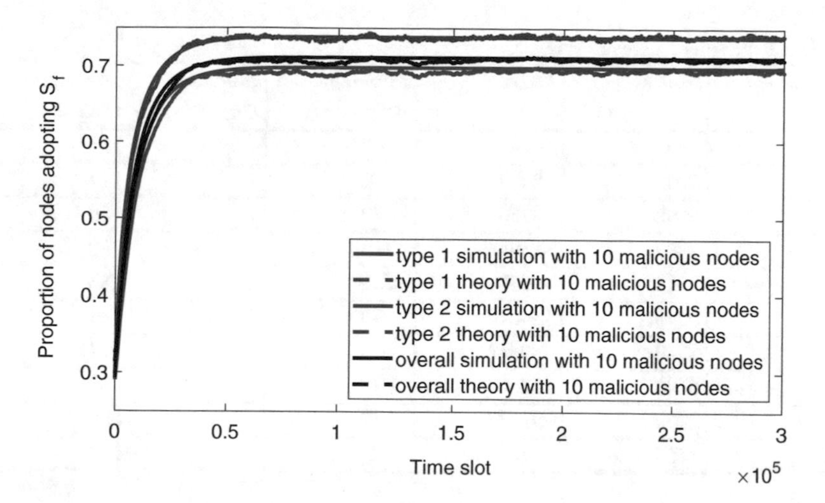

Fig. 3.10 Simulation results for the Erdős-Rényi random network

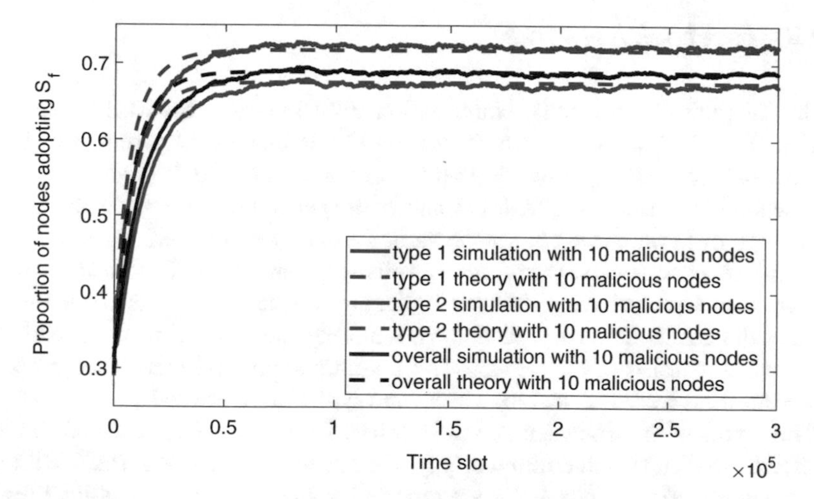

Fig. 3.11 Simulation results for the Barabási-Albert scale-free network

where $\bar{k}$ is the average degree of network, which is set to be 25. For the Barabási-Albert scale-free (BA) network, the degree follows a power law distribution, i.e.,

$$\lambda_{BA}(k) \propto k^{-\xi}. \tag{3.23}$$

In our simulation, we assume that $\xi = 3$. We can see that all the simulation results are consistent with theoretic analysis. Note that the small gap in dynamics of Fig. 3.11 is due to the weak dependence between the network state and the network degree, and it is neglected in the diffusion analysis [29].

Table 3.1 Facebook network statistics

Group index	Nodes	Edges	Maximum degree	Minimum degree	Average degree
1	2235	90954	467	1	81
2	2252	84387	670	1	74
3	3898	137567	1972	1	70
4	3482	155043	773	1	89
5	2314	96394	602	1	83
6	3075	124610	473	1	81
7	2920	89912	478	1	61
8	3445	152007	674	1	88
9	2970	97133	349	1	65
10	1659	61050	577	1	73

3.4.2 Facebook Networks

In this subsection, we simulate information diffusion over real-world social networks. To avoid randomness, we choose ten Facebook networks from [42]. Every network indicates a social friendship network consisting of people (nodes) and edges representing friendship ties. Table 3.1 shows the general network statistics of these ten groups, from which we can see the variance of nodes' number, edges and corresponding network density. Users' degrees change from 1 to 1972, with at least 60% of users in each group having less than 100 degrees. Since we can't tell from the data whether the user is rational or malicious, we designate some of them as malicious users, which consists of the top f_{max} users with maximum degree. Then based on the connections we could get type I users and type II users naturally.

The results of the simulated and theoretic evolutionary stable states of 10 Facebook networks under three different payoff matrices $PM1$, $PM2$, $PM3$ with different number of malicious nodes are reported in Fig. 3.12. The theoretical results come from (3.20), while the simulated results come from the simulation of DB strategy update rule over 10 Facebook networks. It can be seen from the figure that the theoretical results are in good agreement with the simulation results on the Facebook network. The ESSs under different payoff matrices are quite different, signifying that payoff matrix is a crucial factor in determining ESSs. It could also be observed that when malicious nodes' number increases from 10 to 20, ESSs rise from approximately 0.02 to 0.03 as well. Note that under payoff matrix $PM3$, or more broadly $u_{nn} > max\{u_{ff}, u_{fn}\}$, ESSs are not equal to zero due to the presence of malicious users. We notice that there are some small gaps in some networks, such as groups 1 and 7. This is because network's degrees vary widely, which could be reflected from a large maximum degree and a small average degree.

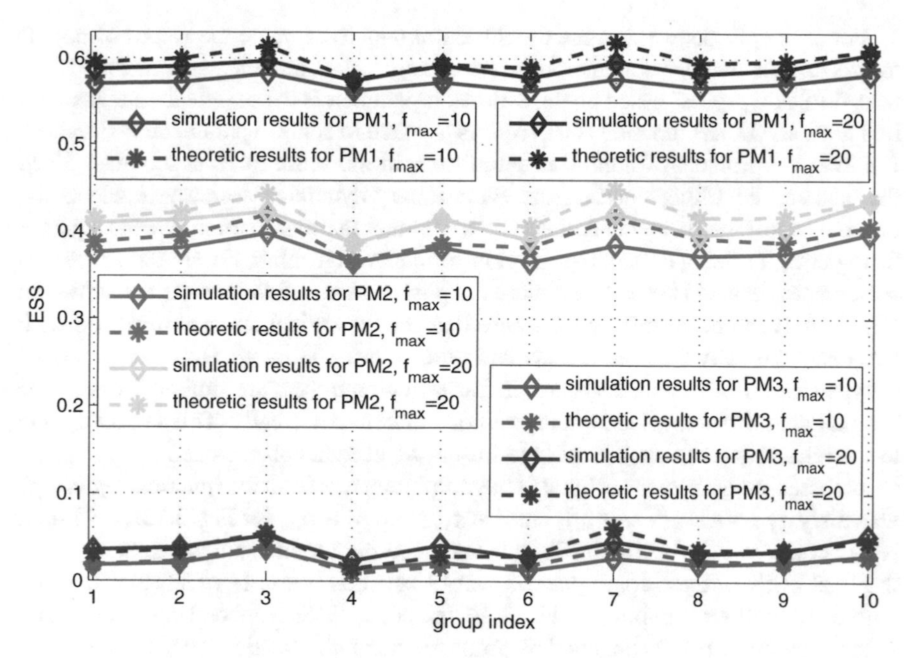

Fig. 3.12 ESSs of real-world Facebook network with different payoff matrices and different number of malicious nodes

3.4.3 Microblog Data Set

In this subsection, the proposed method is further evaluated by real microblog information spreading data set in [43, 44], which involve retweet behaviors in the most popular Chinese microblogging network, Weibo. It comprises 300,000 popular microblog diffusion episodes from Dec. 2010 to Nov. 2012, and each episode includes the content of original microblog, all its retweets and their reposted time stamp. After the original microblog is published, users who are also interested in will retweet and then spread it to more users. Therefore, a diffusion episode can be regarded as the diffusion process of a piece of information. Particularly we choose four diffusion episodes which are negative opinions or misleading information, to test the proposed graphical EGT framework. A total of 36,619 users are covered, whom we assume are all rational. To distinguish type I and type II nodes, we classify users who forward information multiple times in the original data set as type I nodes. This is reasonable, for multiple retweets reflecting user's activeness, and it is also consistent with the definition of type I nodes. By this means the number of type I nodes and type II nodes could be obtained as $M = 1912$ and $N = 34907$. The degree k is set as 100, a typical number of neighbors/friends in social networks. Since the network structure is unknown in the data set, we postulate the network as a constant degree network in which each user has the same degree $k = 100$.

For each episode, we first use the diffusion data to estimate the values of payoff matrix and the number of malicious users by regression, and then generate the theoretic diffusion process based on the estimated parameters to see how it matches with real data. MATLAB function *lsqcurvefit* is invoked to achieve parameter estimation. It solves the problem of nonlinear data fitting in the sense of least squares. Given the data and the fitting function–the evolutionary dynamics, *lsqcurvefit* selects the optimal payoff matrix and malicious users' number in order to minimize the squared fitting error, utilizing default trust-region reflective algorithm. Firstly the entire diffusion episode data is used to estimate the parameters, and the results are shown in Fig. 3.13. It can be seen from the figure that the theoretical curve agrees well with real data especially for overall users and type II users. We notice that the dynamics of type II users and whole users are similar as their numbers are similar. We can also see that the diffusion episode of type I nodes fluctuates greatly. This is mainly due to the relatively small number of type I users M in each episode data.

In these four groups, the estimated payoff parameters from the function *lsqcurvefit* all satisfy $u_{fn} > max\{u_{ff}, u_{nn}\}$, but it is $u_{fn} > u_{ff} > u_{nn}$ for Fig. 3.13a, c while it is $u_{fn} > u_{nn} > u_{ff}$ for Fig. 3.13b, d. Then according to the analysis in Sect. 3.3.3, the final ESSs of these four episodes would between 0 and 1, which is consistent with real situations. As shown in Fig. 3.14, the actual ESSs from real data match well with the theoretic results obtained by proposed method. It could also been seen from Fig. 3.14 that users in group 3 have the most common interests and may be influenced more by malicious nodes as the corresponding ESS is highest. The estimated number of malicious nodes for four groups are 10, 10, 13 and 11, respectively, and it indicates that the population with higher ESS does not necessarily mean more malicious nodes, since the evolutionary dynamics and final ESS are affected by both malicious nodes' number and payoff matrix.

We then evaluate the proposed method in terms of the predictive performance of information diffusion. To this end, only front partial diffusion data is utilized to estimate parameters, and the rest of the data is to be compared with original data. From results in Fig. 3.15 we could observe that the prediction works very well especially for type II nodes and all users. The results of type I nodes are biased, mainly due to the small number of type I nodes and thus the resulting untypical diffusion process.

3.5 Conclusion and Future Work

In this chapter, graphical EGT was utilized to study the harm of malicious nodes in the process of information diffusion over social networks. Firstly rational nodes could be divided into two types according to their connections to malicious nodes: one is type I nodes that is directly connected to the malicious nodes, and the other is type II nodes that is not directly connected to the malicious nodes. Then based on EGT, the evolutionary dynamics and ESSs of type I and type II nodes could be obtained, respectively. To verify our theoretical analysis, we evaluate on synthetic and Facebook networks, as well as real-world microblog data set. Theoretical deriva-

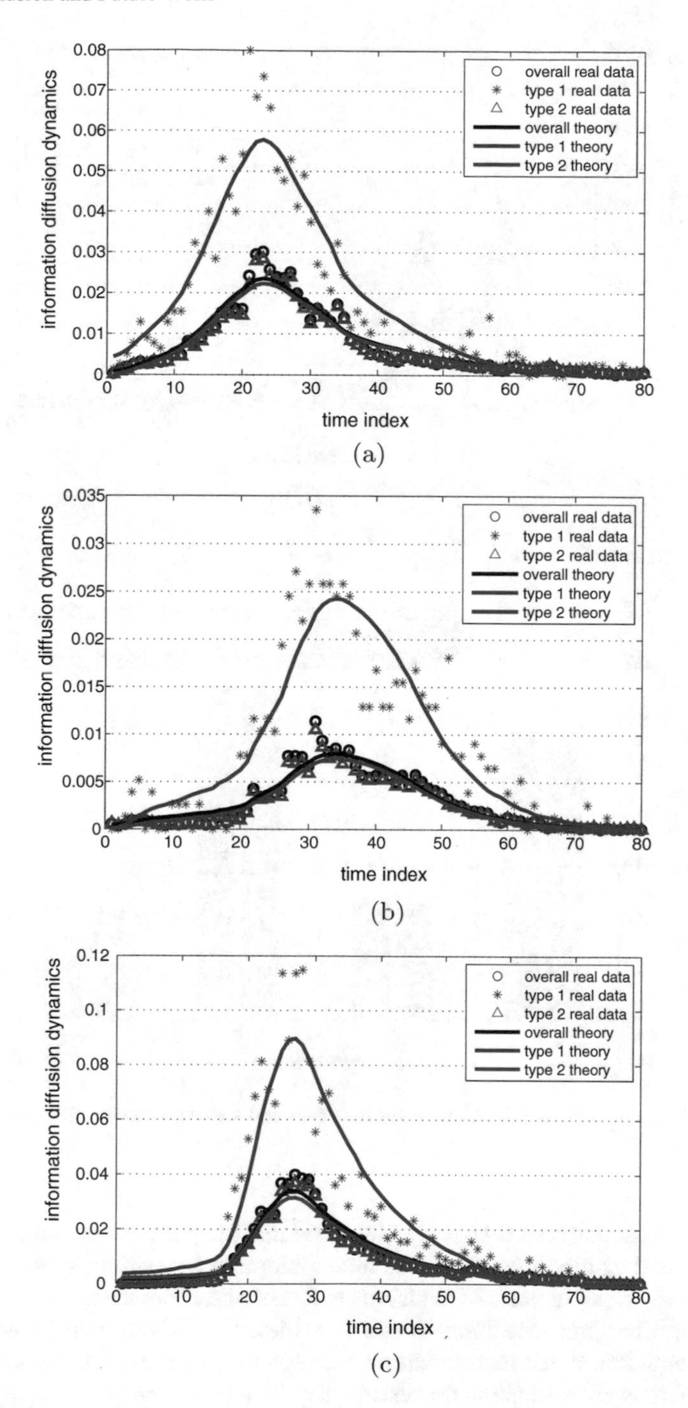

Fig. 3.13 Fitting results of four diffusion episodes

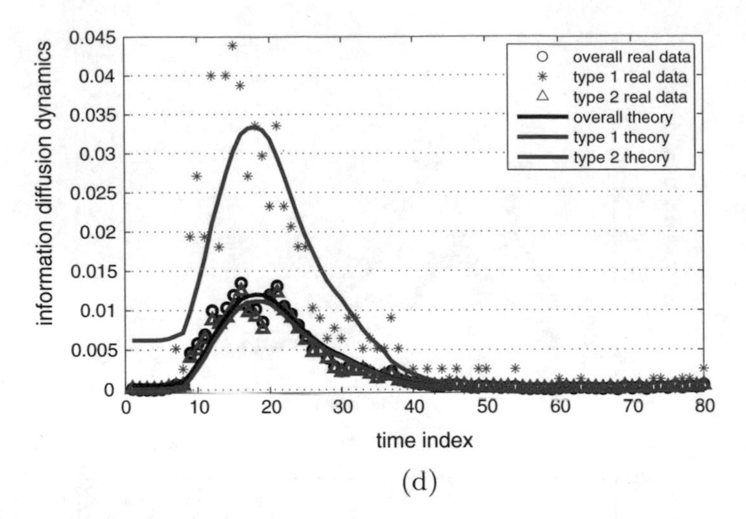

Fig. 3.13 (continued)

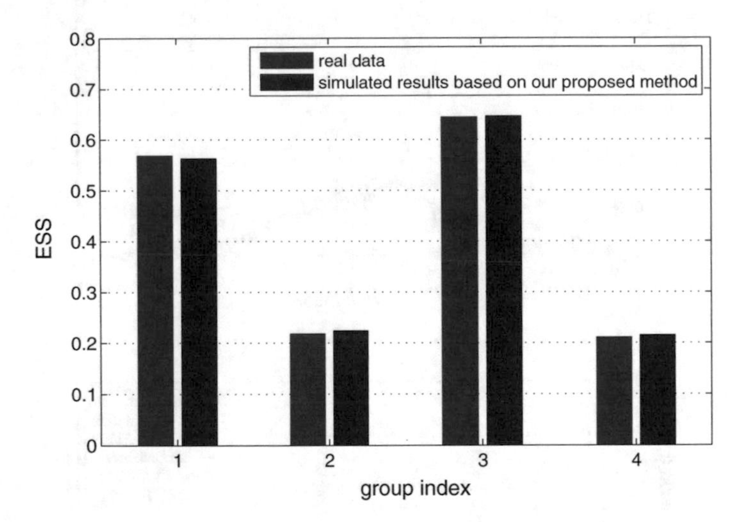

Fig. 3.14 The comparison of ESS between real data and simulated results based on our EGT method

tions and simulation results not only illustrated that the presence of malicious nodes can increase the proportion of rational nodes adopting forwarding strategy, but also quantified the exact impacts of malicious nodes on the entire network.

Although the proposed framework has achieved satisfying performance, there are still some issues that require further research in the future. For instance, in this chapter there is an assumption that users only share their strategies with each other. Nevertheless, in some social networks, users may be familiar with each other and their interests are acquainted. In the circumstances, considering these factors into

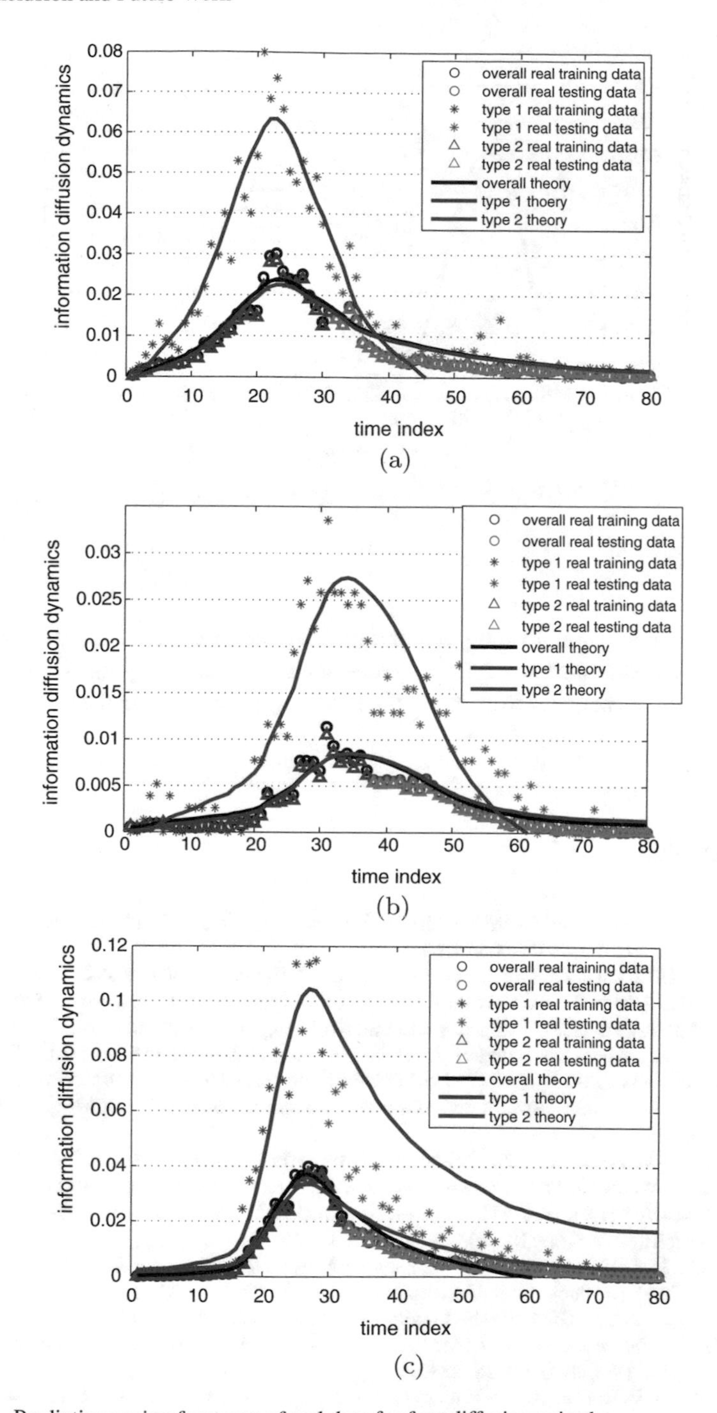

Fig. 3.15 Predictions using front part of real data for four diffusion episodes

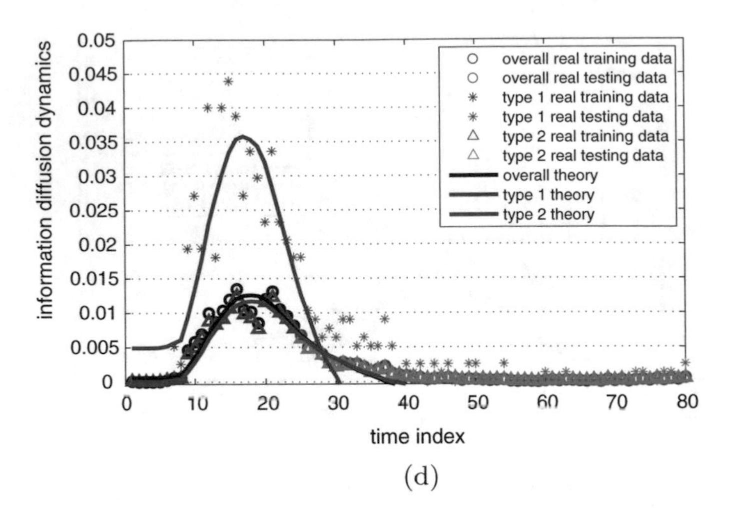

Fig. 3.15 (continued)

the model is more suitable and practical. In addition, users' goals may change over time, so parameters such as payoff matrices need to be time-varying. Therefore, one possible future direction is to consider time-varying parameters in the graphical EGT framework.

References

1. S. Pappas, How big is the internet, really? (2018). [Online]. Available: https://www.livescience.com/54094-how-big-is-the-internet.html
2. B. Marr, How much data do we create every day? The mind-blowing stats everyone should read (2018). [Online]. Available: https://www.forbes.com/sites/bernardmarr/2018/05/21/how-much-data-do-we-create-every-day-the-mind-blowing-stats-everyone-should-read
3. ILOVEYOU, Wikipedia. [Online]. Available: https://en.wikipedia.org/wiki/ILOVEYOU
4. Immanuel, What is the ILOVEYOU virus? How it infects and methods to remove them? (2018). [Online]. Available: https://antivirus.comodo.com/blog/comodo-news/iloveyou-virus-and-its-removal/
5. H. Pinto, J.M. Almeida, M.A. Gonçalves, Using early view patterns to predict the popularity of youtube videos, in *Proceedings of the Sixth ACM International Conference on Web Search and Data Mining*, ser. WSDM '13 (ACM, New York, NY, USA, 2013), pp. 365–374. [Online]. Available: https://doi.org/10.1145/2433396.2433443
6. G. Szabo, B.A. Huberman, Predicting the popularity of online content. Commun. ACM **53**(8), 80–88 (2010). [Online]. Available: https://doi.org/10.1145/1787234.1787254
7. Y. Wang, B. Zheng, On macro and micro exploration of hashtag diffusion in twitter, in *2014 IEEE/ACM International Conference on Advances in Social Networks Analysis and Mining (ASONAM 2014)* (2014), pp. 285–288
8. J. Lehmann, B. Gonçalves, J.J. Ramasco, C. Cattuto, Dynamical classes of collective attention in twitter, in *Proceedings of the 21st International Conference on World Wide Web*, ser. WWW '12 (ACM, New York, NY, USA, 2012), pp. 251–260. [Online]. Available: https://doi.org/10.1145/2187836.2187871

9. Y. Cao, P. Shao, L. Li, Y. Cao, Topic propagation model based on diffusion threshold in blog networks, in *2011 International Conference on Business Computing and Global Informatization* (2011), pp. 539–542
10. X. Lu, Y. Cui, The study of micro-blog information diffusion model based on community structure detection, in *2012 6th International Conference on New Trends in Information Science, Service Science and Data Mining (ISSDM2012)* (2012), pp. 612–616
11. M. Planck, C. Brock, D. Bachman, J. Fenchel, I.L. Pollard, Application of information diffusion in social networks: Verifying the use of web/blog topology entropy as an indicator for real world impact, in *2013 IEEE 2nd Network Science Workshop (NSW)* (2013), pp. 164–167
12. X. Hao, G. Sheng, Z. Yu, L. Juncen, P. Huacan, G. Jun, Predicting information diffusion via matrix factorization based model, in *2014 4th IEEE International Conference on Network Infrastructure and Digital Content* (2014), pp. 257–261
13. L. Alsuwaidan, M. Ykhlef, Information diffusion predictive model using radiation transfer. IEEE Access **5**, 25946–25957 (2017)
14. C. Tsai, J. Lou, W. Lu, S. Lin, Exploiting rank-learning models to predict the diffusion of preferences on social networks, in *2014 IEEE/ACM International Conference on Advances in Social Networks Analysis and Mining (ASONAM 2014)* (2014), pp. 265–272
15. J. Jiang, S. Wen, S. Yu, Y. Xiang, W. Zhou, K-center: an approach on the multi-source identification of information diffusion. IEEE Trans. Inf. Forensics Secur. **10**(12), 2616–2626 (2015)
16. P. Chejara, W. WilfredGodfrey, Machine learning based method to predict influence spread, in *2018 8th International Conference on Cloud Computing, Data Science Engineering (Confluence)* (2018), pp. 268–273
17. J. Yang, N. Zhou, Y. Li, X. Xue, Y. Dong, Y. Xue, J. Li, Opinion-based analysis of structural patterns in online viral diffusion, in *2018 International Conference on Advances in Computing and Communication Engineering (ICACCE)* (2018), pp. 284–289
18. S. Lin, X. Kong, P.S. Yu, Predicting trends in social networks via dynamic activeness model, in *Proceedings of the 22nd ACM International Conference on Conference on Information and Knowledge Management*, ser. CIKM '13 (ACM, New York, NY, USA, 2013), pp. 1661–1666. [Online]. Available: https://doi.org/10.1145/2505515.2505607
19. J.-R. Lee, C.-W. Chung, A new correlation-based information diffusion prediction, in *Proceedings of the 23rd International Conference on World Wide Web*, ser. WWW '14 Companion (ACM, New York, NY, USA, 2014), pp. 793–798. [Online]. Available: https://doi.org/10.1145/2567948.2579241
20. S. Goyal, M. Kearns, Competitive contagion in networks, in *Proceedings of the Forty-fourth Annual ACM Symposium on Theory of Computing*, ser. STOC '12 (ACM, New York, NY, USA, 2012), pp. 759–774. [Online]. Available: https://doi.org/10.1145/2213977.2214046
21. A. Fazeli, A. Jadbabaie, Game theoretic analysis of a strategic model of competitive contagion and product adoption in social networks, in *2012 IEEE 51st IEEE Conference on Decision and Control (CDC)* (2012), pp. 74–79
22. A. Silva, S. Guimarães, W. Meira, Jr., M. Zaki, Profilerank: Finding relevant content and influential users based on information diffusion, in *Proceedings of the 7th Workshop on Social Network Mining and Analysis*, ser. SNAKDD '13 (ACM, New York, NY, USA, 2013), pp. 2:1–2:9. [Online]. Available: https://doi.org/10.1145/2501025.2501033
23. X. Ding, Z. Wu, W. Chen, Y. Liu, Y. Xie, S. Cai, Modeling complex social contagions in big data era, in *2017 IEEE 2nd Advanced Information Technology, Electronic and Automation Control Conference (IAEAC)* (2017), pp. 830–834
24. Z. Jin-lou, L. Zhi-bin, Y. Jian-nan, Modeling of information diffusion based on network dimension-force, in *2011 International Conference on Management Science Engineering 18th Annual Conference Proceedings* (2011), pp. 18–27
25. B. Wang, G. Chen, L. Fu, L. Song, X. Wang, Drimux: dynamic rumor influence minimization with user experience in social networks. IEEE Trans. Knowl. Data Eng. **29**(10), 2168–2181 (2017)
26. C. Jiang, Y. Chen, K.J.R. Liu, Distributed adaptive networks: a graphical evolutionary game-theoretic view. IEEE Trans. Signal Process. **61**(22), 5675–5688 (2013)

27. I.I. Hussein, An individual-based evolutionary dynamics model for networked social behaviors, in *2009 American Control Conference* (2009), pp. 5789–5796
28. C. Jiang, Y. Chen, K.J.R. Liu, Modeling information diffusion dynamics over social networks, in *2014 IEEE International Conference on Acoustics, Speech and Signal Processing (ICASSP)* (2014), pp. 1095–1099
29. C. Jiang, Y. Chen, K.J.R. Liu, Graphical evolutionary game for information diffusion over social networks. IEEE J. Sel. Top. Signal Process. **8**(4), 524–536 (2014)
30. C. Jiang, Y. Chen, K.J.R. Liu, Evolutionary dynamics of information diffusion over social networks. IEEE Trans. Signal Process. **62**(17), 4573–4586 (2014)
31. X. Cao, Y. Chen, C. Jiang, K.J. Ray Liu, Evolutionary information diffusion over heterogeneous social networks. IEEE Trans. Signal Inf. Process. Over Netw. **2**(4), 595–610 (2016)
32. S. Kurihara, The multi agent based information diffusion model for false rumordiffusion analysis (Seoul, Korea, Republic of, 2014), p. 1319, diffusion phenomena; East japan great earthquakes; Information diffusion; Information diffusion models; Multi agent; SIR model; Social networking services; Twitter;. [Online]. Available: https://doi.org/10.1145/2567948.2581449
33. Y. Liu, S. Xu, Detecting rumors through modeling information propagation networks in a social media environment. IEEE Trans. Comput. Soc. Syst. **3**(2), 46–62 (2016)
34. Z. Zhao, X. Chen, Propagation model of derivative rumor considering propagation error and malicious tampering, in *2019 IEEE 4th International Conference on Big Data Analytics (ICBDA)* (2019), pp. 241–245
35. J.-H. Cho, S. Rager, J. O'Donovan, S. Adali, B.D. Horne, Uncertainty-based false information propagation in social networks. Trans. Soc. Comput. **2**(2) (2019) [Online]. Available: https://doi.org/10.1145/3311091
36. A. Josang, J.-H. Cho, F. Chen, Uncertainty characteristics of subjective opinions, in *2018 21st International Conference on Information Fusion (FUSION)* (IEEE, 2018), pp. 1998–2005
37. J. Hofbauer, K. Sigmund, Evolutionary game dynamics. Bull. Am. Math. Soc. **40**, 479–519 (2003)
38. P. Shakarian, P. Roos, A. Johnson, A review of evolutionary graph theory with applications to game theory. Biosystems **107**(2), 66–80 (2012)
39. M.A. Nowak, K. Sigmund, Evolutionary dynamics of biological games. Science **303**(5659), 793–799 (2004). [Online]. Available: https://science.sciencemag.org/content/303/5659/793
40. J.W. Weibull, *Evolutionary Game Theory*, vol. 265 (The MIT Press, 1997)
41. H. Ohtsukia, M.A. Nowak, The replicator equation on graphs. J. Theor. Biol. **243**, 86–97 (2006)
42. R.A. Rossi, N.K. Ahmed, The network data repository with interactive graph analytics and visualization, in *AAAI* (2015). [Online]. Available: http://networkrepository.com
43. J. Zhang, J. Tang, J. Li, Y. Liu, C. Xing, Who influenced you? predicting retweet via social influence locality. ACM Trans. Knowl. Discov. Data **9**(3) (2015). [Online]. Available: https://doi.org/10.1145/2700398
44. J. Zhang, B. Liu, J. Tang, T. Chen, J. Li, Social influence locality for modeling retweeting behaviors, in *Proceedings of the 23rd International Joint Conference on Artificial Intelligence (IJCAI'13)*, pp. 2761–2767

Chapter 4
"Smart" Evolution with Indirect Reciprocity

Abstract When enjoying the convenience of social networks, we are encountering the harm caused by malicious users in social networks as well. In order to reduce their negative effects, it is essential for rational users to carefully screen each connected neighbor to protect themselves from malicious users, implying that establishing a rule for users' interactions in order to mitigate malicious users' influences is required. This chapter introduces the reputation mechanism and proposes a smart evolution model based on evolutionary game theory with indirect reciprocity. The model takes into account both the current reputation and instant incentives in users' decision-making process. After social norms and reputation updating policy are defined, we theoretically analyze the evolutionary dynamics and corresponding evolutionary stable state (ESS) under the proposed scheme. Finally, the validity of the smart evolution model is verified by simulations on synthetic networks, Facebook networks and real-world microblog data set.

Keywords Reputation · Malicious users · Information diffusion · Evolutionary game theory

4.1 Introduction

Nowadays, there is no doubt that the amount of explosive information is shocking and growing faster and faster in the era of big data. According to statistical data from cloud software firm DOMO, in every minute, there are 188,000,000 emails as well as 18,100,000 texts sent worldwide, while 277,777 stories are posted on Instagram, 4,500,000 videos are watched on YouTube, and 511,200 tweets are sent on Twitter per minute. It is predicted by International Data Corporation (IDC) that by 2025, the total amount of digital data globally will increase to 163 zettabytes with growing devices and sensors [1, 2]. As these numbers balloon, it's no surprise that we have access to huge amount of data, most of which are critical today.

These statistics manifest people's interactions and are also pivotal to the marketing activities of many companies. Now people favor surfing on large-scale social networks such as Facebook, Twitter, Instagram and WeChat, to chat with each other and

Y. Chen and H. V. Zhao, *Behavior and Evolutionary Dynamics in Crowd Networks*, Lecture Notes in Social Networks, https://doi.org/10.1007/978-981-15-7160-2_4

share all kinds of information. As a result, information in various forms is published or transmitted by many individuals or companies, even if consisting of both positive ones and negative ones. Some network users would diffuse bad news for their particular purposes regardless of other users' benefits. What's worse is that some malicious users might spread harmful information, such as viruses, which brings serious consequences and immeasurable damage to society. Therefore, rational users, who make their decisions in a reasonable manner, need to keep themselves from the harmful effects of malicious users. But how can rational users distinguish unfriendly or malicious users? How can rational users update their strategies to alleviate the effects of malicious users? What will rational users' resistance lead to? All these issues need to be addressed.

There are many factors acting on the intricate process of rational users' decision-making, such as the authenticity of information, the duration of neighbors' strategy, individuals' personal preference and decisions of others. For instance, if a neighbor node spreads false information, or deliberately reposts a piece of information for a long time, or disseminates various information that is not of interest to the center node, he/she may be distrusted by the center user and might even be blocked. Since one's decision-making involves the interactions between multiple decision makers, game theory is a very appropriate mathematical modeling tool [3]. In social networks, under the circumstance that not everyone is rational and not all information is obtainable to users, evolutionary game theory (EGT) is more suitable compared with traditional game theory. EGT imagines that the game is played over and over again by socially conditioned players who are "pre-programmed" to some behaviors–formally a strategy in the game, and the evolutionary selection process operates over time on the population distribution of behaviors, which is more close to the reality [4]. With EGT, we can consider all of the above factors into the model for setting users' interaction policy and updating rules. It can promote understanding and prediction of users' behavior, while reducing the harmful effects of malicious users.

In the literature, many scholars have utilize evolutionary game theory to model the interactions over social networks. Hussein in [5] researched on individual behavioral choices by evolutionary dynamics over a graph of connected individuals, emphasizing the individuality of the nodes. The adaptive network was investigated from the game theoretic perspective and the distributed adaptive filtering problem was formulated as a graphical evolutionary game in [6]. Authors proposed an evolutionary game theoretic framework in [7, 8] to model the dynamic information diffusion process in social networks, while [7] paid more attention to the final stable state and [8] stressed on evolutionary dynamics. In [9], by modeling the interactions among the heterogeneous users as a graphical evolutionary game, Cao et al. derived the evolutionary dynamics and the evolutionarily stable states (ESSs) of the information diffusion. Also the study of irrational behaviors' impacts on information propagation under the evolutionary game-theoretic framework was carried out in [10]. In [7–10], the information diffusion process was described and predicted in detail, and simulation results show that the theoretical analysis is in good agreement with real conditions. However, the researches on the evolutionary dynamics and evolutionary stable strategies of networks with some irrational users were not in-depth, nor had them considered how to quantify the adverse effects of irrational users based on EGT.

The work of this chapter is more correlative to bad influences minimization, which is opposed to influence maximization. Influence maximization is one of the important areas of social network analysis since it contributes to discover influential entities, or in other words seeds, then make the information ultimately spread the most influence in the network. The influence maximization problem was studied in many works and various seed selection algorithms were proposed [11–21], such as a simple greedy adaptive seeding strategy and an efficient heuristic algorithm in [12], a centrality-based edge activation probability evaluation method in [11], a polling-based randomized algorithm in [13], the dynamic probability based genetic approach using topic affinity propagation (TAP) method in [14], and the centrality-based methods in [15]. To make the influence maximization efficient especially in large-scale network, authors in [16] proposed a divide-and-conquer strategy with parallel computing mechanism, while in [17] a linear time iterative approach was proposed. Zhang et al. in [18] mapped a set of networks into a single one via lossless and lossy coupling schemes to get minimal seeds and save cost. Authors in [19] utilized a greedy algorithm with a constant approximation ratio to derive minimal seeds as well. In [20, 21], the ways to find both the best set of seeds and the right timing to perform these seedings are figured out. Unlike influence maximization, in order to minimize the negative impacts of bad information such as rumors and viruses in social networks, researchers studied the problem of impact blocking maximization (IBM). A community based algorithm called FC-IBM algorithm using fuzzy clustering and centrality measures was proposed in [22] to find a good candidate subset of nodes for diffusion of positive information to compete against negative information. Authors in [23] achieved the IBM by identifying a minimal subset of nodes and then removing all the nodes in this subset as well as their incoming and outgoing edges from the network. In [24, 25], a certain subset of individuals were found to block propagation. However, these works resolved this problem from a macro perspective, i.e., they paid more attention to the graphical structure of the network, and ignored the important role of the individuals in minimizing the adverse effects.

Different from IBM methods, in order to decline the negative influences in social network, we focus on the mechanism of individuals' interactions, or specifically how rational users protect themselves from hazard caused by malicious users. Generally users in the network tend to cooperate, by which users could know more information about their neighbors. Thus indirect reciprocity is introduced in the analysis to stimulate the users' incentive to play cooperatively. Indirect reciprocity is a key mechanism for the evolution of human cooperation and has recently drawn a lot of attentions in the area of social science and evolutionary biology [26, 27]. The key concept of indirect reciprocity is "I help you not because you have helped me but because you have helped others". Many works have applied indirect reciprocity into their formulation. In [28] a cooperation stimulation scheme based on indirect reciprocity was proposed, for the scenario where interactions between any pair of players are finite in cognitive networks. Xiao et al. proposed a security system that applied the indirect reciprocity principle to combat attacks in wireless networks [29].

Authors in [30] discussed a reputation-based incentive framework, where the data sharing stimulation problem was modeled as an indirect reciprocity game.

In this chapter, a smart evolution model of information diffusion is discussed, integrated with evolutionary game theory and indirect reciprocity. This model puts more emphasis on personal decisions and motivations, and its objective is to cut down the negative impacts of malicious users over the whole network. To distinguish the reliability of each user, a reputation update policy is established according to the social norm. With indirect reciprocity principle, each node in the network checks the actions of neighbors, updates reputation values of neighbors, and broadcasts the new reputation to entire network through the gossip channel, i.e., all nodes share reputation with others rather than updating independently. Based on this, the theoretical analysis of smart evolution model is carried out, and corresponding evolutionary dynamics and ESSs are obtained. Finally, we conduct simulations on both synthetic data and real-world data to validate theoretical analysis. The results show that compared with the traditional evolution model in which rational users treat people around them equally, the smart evolution model can efficiently alleviate malicious users' negative impacts.

4.2 System Model

In this section, the smart evolution model for information diffusion is introduced in detail. We first elaborate on social network model in Sect. 4.2.1, and then define the payoffs and fitness of EGT framework in Sect. 4.2.2. In Sect. 4.2.3, we discuss the strategy updating rules for all users. Finally, the social norm and reputation updating policy of rational users and malicious users are described in Sects. 4.2.4 and 4.2.5, respectively.

4.2.1 Social Network Model

As shown in Fig. 4.1, the social network could be modeled as an undirected graph, with nodes representing users and edges representing their mutual connections. In the graph, users who have reasonable interactions with others and will not intentionally post negative or malicious information are defined as rational nodes. Those who spread useless, false or even harmful information on purpose so as to achieve some certain purposes, are defined as malicious nodes. By connecting with people as much as possible, they plan to expand their adverse effects.

Due to the existence of malicious nodes, rational nodes could be further categorized into two types: one is type I nodes, which connect directly to malicious nodes, while the other is type II nodes, which connect indirectly to malicious nodes. The numbers of type I nodes, type II nodes and malicious nodes are assumed to be M, N and a_{max}, respectively. Thus there are totally $M + N$ rational nodes in the network.

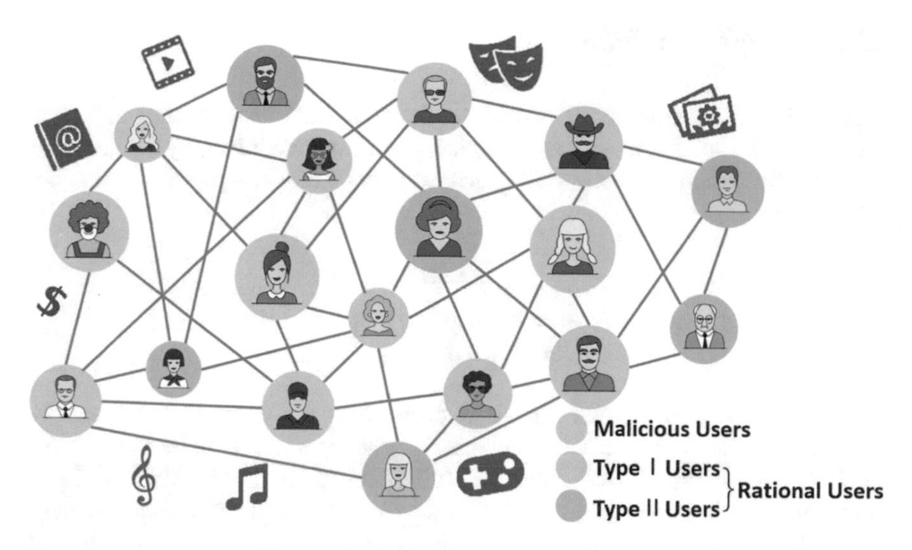

Fig. 4.1 Network structure

For a type I rational node, it has k rational neighbors, which obeys the distribution $\lambda(k)$, and a malicious neighbors which obeys $\mu(a)$. The distribution $\lambda(k)$ means that when randomly choosing a rational node, it has k adjacent rational nodes with the probability $\lambda(k)$, and the implication for $\mu(a)$ is similar. While for a type II rational node, it only has k rational neighbors on account of the absence of malicious nodes. In general, rational users would not actively establish relationships (links) with completely unfamiliar users. Hence, malicious users have to initially contact rational users on their own initiative, some of whom are willing to bc connected and are exactly type I nodes. We assume that the connections between malicious users and type I users are established at the first moment of information diffusion.

During the information diffusion, when a user receives one piece of information, he/she only has two choices: forwarding and not forwarding. In such a case, we define the strategy set as $S = \{0, 1\}$, where 0 indicates forwarding the information and 1 indicates not forwarding, and a user's strategy $s \in S$. To grasp the condition of every moment in propagation, we use population state p_f and local population state p_{f1}, p_{f2}, which are defined as percentages of users choosing strategy $s = 0$ for all rational users, type I users and type II users, respectively. The relationship between p_f, p_{f1} and p_{f2} is

$$p_f = \frac{M}{M+N} p_{f1} + \frac{N}{M+N} p_{f2}. \tag{4.1}$$

And the expected variations of per unit time $\dot{p}_f$, $\dot{p}_{f1}$ and $\dot{p}_{f2}$ are defined as the dynamics of p_f, p_{f1} and p_{f2}, respectively.

4.2.2 Payoffs and Fitness

In the evolution, strategies of individuals are tested under "the rules of game", which involve payoffs and the fitness, and they are generally determined by the characteristics of information and network's structure. Specifically, the payoff matrix could be defined as

$$\begin{array}{cc} & \begin{array}{cc} s=0 & s=1 \end{array} \\ \begin{array}{c} s=0 \\ s=1 \end{array} & \begin{pmatrix} u_{ff} & u_{fn} \\ u_{nf} & u_{nn} \end{pmatrix}, \end{array} \tag{4.2}$$

where u_{ff} and u_{nn} are payoffs when both users adopt $s = 0$ or $s = 1$, respectively, and $u_{fn} = u_{nf}$ is the payoff when one user's strategy is $s = 0$ while the other's is $s = 1$. Apparently the matrix is symmetric. Moreover, the payoff is assumed to be normalized within interval (0, 1), i.e., $0 < u_{ff}, u_{fn}, u_{nn} < 1$. The payoff matrix is universal for all users in the network under the assumption that users couldn't identify each other's type and they only exchange information about strategies and reputation values when interacting.

Based on the payoff matrix, the fitness of a user π, also known as the utility of a user, is defined as

$$\pi = (1 - \alpha)B + \alpha U, \tag{4.3}$$

in which B is the baseline fitness and is often set as 1, U is user's payoff from payoff matrix, and $\alpha \in (0, 1)$ is the selection intensity, i.e., the relative contribution of the current payoff to fitness. We assume that $\alpha \to 0$, corresponding to the case of weak selection [7–9]. The assumption is reasonable because it means that people doubt at some level initially towards new received information, so current payoff should not be the main part in the updating of fitness.

4.2.3 Strategy Updating Rules

Generally, a rational node updates its strategy based on many factors such as the user's own interests and the behaviors of neighbors. For example, when most of the neighbors forward this information, or when the fitness of forwarding is higher than that of non-forwarding, the center node is more likely to forward the information as well. In this chapter, we specify the Death-birth (DB) updating rule as the regulation for rational nodes. The updating process is as follows: a random player is chosen to abandon his/her current strategy (Death process). Then, the chosen player adopts one of his/her neighbors' strategies with the probability being proportional to their fitness (Birth process). The analyses of other strategy update rules such as Birth-death (BD) update rule and imitation (IM) update rule are equivalent particularly when the network degree k is sufficiently large [8, 31].

For malicious nodes, their decision-making of strategies do not conform to normal rules due to their evil motives. They can choose any strategy at any time to achieve their goals. Here, we consider the scenario that malicious users spread false or deceptive information, and to make these information more popular they constantly disseminate, inform their rational neighbors so as to influence more rational nodes to the greatest extent. Therefore, the strategy of malicious node is assumed to be $s = 0$ all the time.

4.2.4 *Social Norm*

The notion of reputation is the key concept of indirect reciprocity. The reputation mechanism evaluates players' behavior through social norms, by which users can update their neighbors' reputation and estimate their reliability, and then the negative impacts of malicious nodes could be greatly reduced. In this subsection, we elaborate on social norms for malicious and rational users respectively.

4.2.4.1 Social Norm for Malicious Nodes

Note the assumption that the connections between malicious nodes and rational nodes are established at first, so malicious neighbors are totally new friends to type I nodes. For unfamiliar neighbors, it is necessary to build a different social norm to evaluate their reliability. According to experiences in real life, there are three main factors that influence the reputation of malicious nodes. The first one is the authenticity of information, i.e., whether it is true if the user adopted forwarding strategy before. False information will lead to a low instant reputation and thus reduce the user's reputation. The second factor is the duration of same strategy. For instance, if a user always adopts $s = 0$, it may be regarded as malicious marketing. The longer the duration, the lower the user's reputation. And the third factor is the consistency of the neighbor and the center node. If center node has the same strategy with the neighbor, he/she may feel that they have similar interests and hobbies. In such a case, the prior factor could be ignored.

Based on these three factors, we define the social norm of malicious nodes, Q_M, as a 2×2 matrix as follows

$$Q_M = \begin{array}{c} \\ s_r{=}0 \\ s_r{=}1 \end{array} \begin{array}{c} \begin{array}{cc} s_m{=}0 & s_m{=}1 \end{array} \\ \begin{pmatrix} 1+d & c \\ c & 1+d \end{pmatrix} \end{array}, \tag{4.4}$$

where s_m and s_r are the strategies of malicious node and corresponding connected type I node. The d and c are defined as

$$d = s_m + \lambda, \tag{4.5}$$

and

$$c = \begin{cases} 1 + d, & n \leq n_a \\ 1 - d, & n > n_a \end{cases}, \tag{4.6}$$

where n_a means the maximal duration that a rational user could stand. If the duration of same strategy n exceeds this threshold, the instant reputation would be low and thus the reputation would decrease according to its updating policy. λ means the trueness of information. $\lambda = 1$ if the information is true and otherwise $\lambda = 0$. We assume that it takes some time to clarify whether the information forwarded before is true or not, which is defined as t_0. In other words, only after t_0 time slots can other users know the authenticity of information that center user forwards.

4.2.4.2 Social Norm for Rational Nodes

Unlike the situation for malicious nodes, rational nodes are generally familiar with each other, thus they know each other's hobbies and don't mind their strategies. In such a case, we assume that only the authenticity of information affects the reputation of rational nodes. The social norm for rational users is defined as

$$\boldsymbol{Q_R} = \begin{array}{c} \\ s_r{=}0 \\ s_r{=}1 \end{array} \begin{array}{c} \begin{array}{cc} s_r{=}0 & s_r{=}1 \end{array} \\ \begin{pmatrix} 1+d & 1+d \\ 1+d & 1+d \end{pmatrix} \end{array}, \tag{4.7}$$

where d is

$$d = s_r + \lambda, \tag{4.8}$$

with s_r being the strategy of rational node.

4.2.5 Reputation Updating Policy

To measure the probability of a user with a high or low reputation, we define the reputation distribution $\boldsymbol{r} = [r_l, r_h]$ with $r_l + r_h = 1$, where r_l indicates the probability of user being assigned a low reputation and r_h indicates the probability with a high reputation. At time slot t, every user i has a unique reputation distribution, which is denoted as $\boldsymbol{r}_r^i(t)$ for a rational user and $\boldsymbol{r}_m^i(t)$ for a malicious user. At first, users aren't able to discriminate others' credibility, so the reputation distribution of all nodes is set as [0, 1] at the first time slot.

Figure 4.2 shows the reputation updating policy that is applicable for both rational and malicious nodes, incorporating social norms $\boldsymbol{Q_R}$ and $\boldsymbol{Q_M}$ discussed in the

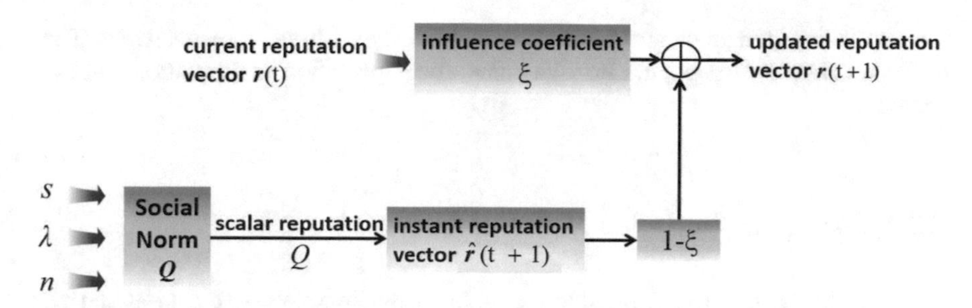

Fig. 4.2 The reputation updating model

previous subsection. According to social norms, the scalar reputation Q is obtained, which could be 0,1,or 2. If $Q = 2$, it means that this node is reliable and a high instant reputation should be assigned. If $Q = 1$, it means that this node couldn't be trusted at present, and a low instant reputation should be assigned. If $Q = 0$, it represents that the reliability of this node could not be confirmed and the reputation doesn't change. Then the instant reputation distribution for rational nodes and malicious nodes in time slot $(t + 1)$ can be written as

$$\hat{\boldsymbol{r}}_r^i(t+1) = \boldsymbol{e}_{Q_{R_{s_{r1},s_{r2}}}}, \tag{4.9}$$

and

$$\hat{\boldsymbol{r}}_m^i(t+1) = \boldsymbol{e}_{Q_{M_{s_r,s_m}}}, \tag{4.10}$$

where $\boldsymbol{e}_2 = [0, 1]$, $\boldsymbol{e}_1 = [1, 0]$, $\boldsymbol{e}_0 = \boldsymbol{r}_r^i(t)$ in (4.9) and $\boldsymbol{e}_0 = \boldsymbol{r}_m^i(t)$ in (4.10).

With the instant reputation distribution, user i updates the reputation distribution at time index $(t + 1)$ using a linear combination of reputation distribution of previous time index and the current instant reputation distribution with a weight ξ as

$$\boldsymbol{r}_r^i(t+1) = \xi \boldsymbol{r}_r^i(t) + (1-\xi)\hat{\boldsymbol{r}}_r^i(t+1), \tag{4.11}$$

and

$$\boldsymbol{r}_m^i(t+1) = \xi \boldsymbol{r}_m^i(t) + (1-\xi)\hat{\boldsymbol{r}}_m^i(t+1). \tag{4.12}$$

4.3 Information Diffusion Model Based on Indirect Reciprocity

In this section, we analyze the information diffusion process based on indirect reciprocity principle. In indirect reciprocity game, reputation values of all users are shared over whole social network via gossip channel, which would help users know each other more comprehensively and accelerate the process to final ESSs. We first study the population dynamics and ESSs of type I and type II nodes respectively. Then to further simplify the population dynamics, the expectation of reputation values is solved.

4.3.1 Analysis for Type I Rational Users

Based on the fundamental form of fitness in (4.3), the fitness for a type I rational user adopting strategy $s = 0$ and $s = 1$ in time slot t could be respectively written as

$$\pi_f(t) = 1 - \alpha + \alpha \left(R_1 u_{ff} + R_2 u_{fn} \right), \tag{4.13}$$

and

$$\pi_n(t) = 1 - \alpha + \alpha \left(R_1 u_{fn} + R_2 u_{nn} \right), \tag{4.14}$$

where $R_1 = \sum_{i=1}^{k_f} r_{rh}^i(t) + \sum_{i=1}^{a} r_{mh}^i(t)$, $R_2 = \sum_{i=1}^{k-k_f} r_{rh}^i(t)$, k denotes the number of neighbor nodes, and k_f denotes the number of neighbors with strategy $s = 0$.

In each round of the DB update, one of the $M + N$ users will be selected to update his/her strategy randomly. The proportion of type I users with strategy $s = 0$ among all rational users is $p_{f1}M/(M + N)$. If one of these users is chosen, he/she may change the current strategy to $s = 1$ with the probability $P_{f\to n}^t$ denoted as

$$P_{f\to n}^t = \frac{R_2 \pi_n(t)}{R_1 \pi_f(t) + R_2 \pi_n(t)}. \tag{4.15}$$

If strategy transition happens, it would make p_{f1} decrease by $1/M$.

On the contrary, p_{f1} may increase by $1/M$ when a type I rational user with strategy $s = 1$ is chosen and he/she decide to change the strategy to $s = 0$ as well. The probability of being chosen is $(1 - p_{f1})M/(M + N)$, and the probability of strategy transition could be denoted as

$$P_{n\to f}^t = \frac{R_1 \pi_f(t)}{R_1 \pi_f(t) + R_2 \pi_n(t)}. \tag{4.16}$$

Combining these two scenarios, the population dynamics for type I node are derived as

$$\begin{aligned}
\dot{p}_{f1} &= \mathbf{E}\left[\frac{p_{f1}M}{M+N}\cdot P^t_{f\to n}\cdot\left(-\frac{1}{M}\right)+\frac{(1-p_{f1})M}{M+N}\cdot P^t_{n\to f}\cdot\frac{1}{M}\right]\\
&= \frac{1}{M+N}\mathbf{E}\left\{\frac{R_1}{R_1+R_2}-p_{f1}+\alpha\frac{R_1R_2\left[R_1u_{ff}+(R_2-R_1)u_{fn}-R_2u_{nn}\right]}{(R_1+R_2)^2}\right\}\\
&= \frac{1}{M+N}\mathbf{E}\left\{\frac{kp_f\cdot r^i_{rh}(t)+a\cdot r^i_{mh}(t)}{k\cdot r^i_{rh}(t)+a\cdot r^i_{mh}(t)}-p_{f1}+\alpha k\cdot\right.\\
&\frac{(k-1)\left[-\Phi(k-2)p_f^3+(k\Phi-3\Phi-k\Phi_n)p_f^2+(\Phi+k\Phi_n)p_f\right]\left[r^i_{rh}(t)\right]^3+}{k^2\left[r^i_{rh}(t)\right]^2+a^2\left[r^i_{mh}(t)\right]^2+2kar^i_{rh}(t)r^i_{mh}(t)}\\
&\frac{\left[-a(k-1)(2\Phi+\Phi_n)p_f^2+a(2k\Phi-2\Phi-\Phi_n)p_f+ak\Phi_n\right]\left[r^i_{rh}(t)\right]^2r^i_{mh}(t)+}{}\\
&\left.\frac{a^2(\Phi+\Phi_n)(1-p_f)r^i_{rh}(t)\left[r^i_{mh}(t)\right]^2}{}\right\},
\end{aligned} \tag{4.17}$$

where $\Phi = u_{ff} - 2u_{fn} + u_{nn}$ and $\Phi_n = u_{fn} - u_{nn}$. During the derivation of (4.17), we utilize the moments of binomial distribution: $\mathbf{E}(k_f) = kp_f, \mathbf{E}(k_f^2) = k(k-1)p_f^2 + kp_f, \mathbf{E}(k_f^3) = k(k-1)(k-2)p_f^3 + 3k(k-1)p_f^2 + kp_f$, since neighbors' strategies could be regarded as a Bernoulli sequence with mean being p_f. It's reasonable because the scale of social network is sufficiently large.

4.3.2 *Analysis for Type II Rational Users*

When considering neighbors of type II users, malicious users are out of consideration and the number is k rather than $(k+f)$. Therefore, based on (4.3) we could get the fitness for type II rational user in time slot t as

$$\pi_f(t) = 1-\alpha+\alpha\left(R_3u_{ff}+R_2u_{fn}\right), \tag{4.18}$$

and

$$\pi_n(t) = 1-\alpha+\alpha\left(R_3u_{fn}+R_2u_{nn}\right). \tag{4.19}$$

where $R_3 = \sum_{i=1}^{k_f} r^i_{rh}(t)$.

Similarly, a type II node is chosen with a specific probability. He/She may change his/her current strategy with transition probability

$$P^t_{f\to n} = \frac{R_2\pi_n(t)}{R_3\pi_f(t)+R_2\pi_n(t)}, \tag{4.20}$$

and

$$P_{n\to f}^{t} = \frac{R_3\pi_f(t)}{R_3\pi_f(t) + R_2\pi_n(t)}, \tag{4.21}$$

which would lead to the fluctuation of p_{f2}.

Then, we could derive the population dynamics of type II nodes $\dot{p}_{f2}$ in a similar way as

$$\begin{aligned}
\dot{p}_{f2} &= \mathbf{E}\left[\frac{p_{f2}N}{M+N} \cdot P_{f\to n}^{t} \cdot \left(-\frac{1}{N}\right) + \frac{(1-p_{f2})N}{M+N} \cdot P_{n\to f}^{t} \cdot \frac{1}{N}\right] \\
&= \frac{1}{M+N}\mathbf{E}\left\{\frac{R_3}{R_2+R_3} - p_{f2}! + \alpha \frac{R_2R_3\left[R_3u_{ff} + (R_2-R_3)u_{fn} - R_2u_{nn}\right]}{(R_2+R_3)^2}\right\} \\
&= \frac{1}{M+N}\mathbf{E}\left\{\frac{kp_f \cdot r_{rh}^{i}(t)}{k \cdot r_{rh}^{i}}(t) - p_{f2} + \alpha(k-1)p_f \cdot \right. \\
&\left. \frac{\left[-\Phi(k-2)p_f^2 + (k\Phi - 3\Phi - k\Phi_n)p_f + \Phi + k\Phi_n\right]r_{rh}^{i}(t)}{k}\right\}.
\end{aligned} \tag{4.22}$$

Compared with (4.17), (4.22) appears much more precise, which is due to the absence of malicious nodes in the calculation of neighbors.

4.3.3 *Expectation of Reputation Values*

Now we obtain the population dynamics for type I and type II rational users in (4.17) and (4.22), where the expectation is on the reputation distribution. So in this subsection, we solve the expectation of reputation values for malicious nodes and rational nodes, respectively.

4.3.3.1 Expectation of Malicious Nodes' Reputation

Firstly we consider the reputation of malicious nodes. Previously only the probability with a high reputation is utilized in the calculation of dynamics, i.e., $r_{m_h}^{i}(t)$, so we specify its updating rule $r_{m_h}^{i}(t+1) = \xi r_{m_h}^{i}(t) + (1-\xi)\hat{r}_{m_h}^{i}(t+1)$. From the rule, it could be seen that the current reputation is updated based on prior reputation. Then by iteration the relationship between current reputation and initial reputation can be derived as

$$r_{m_h}^{i}(t+1) = \xi^t r_{m_h}^{i}(1) + (1-\xi)\sum_{j=2}^{t+1}\xi^{t+1-j}\hat{r}_{m_h}^{i}(j). \tag{4.23}$$

Then, the $\left[r_{m_h}^{i}(t+1)\right]^2$ is

$$\left[r_{m_h}^i(t+1)\right]^2 = \xi^{2t}\left[r_{m_h}^i(1)\right]^2 + (1-\xi)^2\left[\sum_{j=2}^{t+1}\xi^{t+1-j}\hat{r}_{m_h}^i(j)\right]^2$$
$$+2r_{m_h}^i(1)(1-\xi)\xi^t\sum_{j=2}^{t+1}\xi^{t+1-j}\hat{r}_{m_h}^i(j). \quad (4.24)$$

To get the expectation of (4.23) and (4.24), we simplify them as

$$\mathbf{E}\left\{r_{m_h}^i(t+1)\right\} = \xi^t r_{m_h}^i(1) + \left(1-\xi^t\right)\mathbf{E}\left\{\hat{r}_{m_h}^i(j)\right\} \quad (4.25)$$

and

$$\mathbf{E}\left\{\left[r_{m_h}^i(t+1)\right]^2\right\} = \xi^{2t}\left[r_{m_h}^i(1)\right]^2 + (1-\xi^t)^2\mathbf{E}\left\{\left[\hat{r}_{m_h}^i(j)\right]^2\right\}$$
$$+2r_{m_h}^i(1)\xi^t(1-\xi^t)\mathbf{E}\left\{\hat{r}_{m_h}^i(j)\right\}, \quad (4.26)$$

from which we could see that the expectation is on $\hat{r}_{m_h}^i(j)$ and $\left[\hat{r}_{m_h}^i(j)\right]^2$.

When reputation is shared over the network, i.e., there is only one reputation value for every user, a user's reputation is updated by all his/her neighbors. During each DB updating round, there is only one randomly selected node, so rational users would not update his/her neighbors' reputation in every time slot, which means that the time slot j in (4.23) is not continuous. In the choosing process, the probability of a type I node being selected is $M/(M+N)$. For type I nodes, the numbers of malicious neighbors are different. Averagely, it could be considered as $\bar{a}$, calculated by $\bar{a} = \sum_{a=1}^{\infty}\mu(a)a$. Hence, the reputation of a malicious node could be considered to be updated every $a_{max}(M+N)/(\bar{a}M)$ time slots. In the following, $\lceil\rceil$ stands for ceiling function.

Generally speaking, our patience for the invariance of strategy n_a is usually longer than the time we need to know the authenticity of information t_0, since we tend to forgive others initially. Therefore, we assume that $t_0 \le n_a$. The probability that a rational node connected to a malicious node uses $s=0$ can be regarded as p_{f1}. In such a case, according to the social norm, the expectation of instant reputation for malicious nodes could be divided into following two situations:

- When $t \le t_0$, rational users have not known the trueness of information yet. As a result, $\hat{r}_{m_h}^i(t) = 1$ and $\left[\hat{r}_{m_h}^i(t)\right]^2 = 1$.
- When $t > t_0$, $\hat{r}_{m_h}^i(t)$ could be regarded as a Bernoulli sequence. Then the expectation of $\hat{r}_{m_h}^i(t)$ and $\left[\hat{r}_{m_h}^i(t)\right]^2$ is p_2, which is the probability that the information forwarded by malicious user is true.

The probability that a rational node connected to a malicious node adopts $s=0$ can be regarded as $(1-p_{f1})$. Under the circumstance, the expectation of instant reputation value for malicious nodes could be divided into following three situations:

- When $t \le t_0, n \le n_a$, it's the same as before, that is $\hat{r}^i_{m_h}(t) = 1$ and $\left[\hat{r}^i_{m_h}(t)\right]^2 = 1$.
- When $t > t_0, n \le n_a$, $\hat{r}^i_{m_h}(t)$ could be regarded as a Bernoulli sequence. Then $\mathbf{E}\left\{\hat{r}^i_{m_h}(t)\right\} = \mathbf{E}\left\{\left[\hat{r}^i_{m_h}(t)\right]^2\right\} = p_2$.
- When $t > t_0, n > n_a$, $\hat{r}^i_{m_h}(t)$ could only be 0 with probability $(1 - p_2)$, or reputation of previous time slot, i.e., $r^i_{m_h}(t-1)$, with probability p_2. If $\hat{r}^i_{m_h}(t) = r^i_{m_h}(t-1)$, $r^i_{m_h}(t)$ would not change and thus the iteration could be omitted. So in the remaining $\lceil (t - n_a)/(M+N) \rceil (1 - p_2)$ iterations, the expectation of $\hat{r}^i_{m_h}(t)$ and $\left[\hat{r}^i_{m_h}(t)\right]^2$ is 0.

In summary, the final expectation for malicious users' reputation in (4.25) and (4.26) is derived as follows

$$
\begin{aligned}
&\mathbf{E}\left\{r^i_{m_h}(t+1)\right\} = \\
&\begin{cases}
1, & t \le t_0, \\
\xi^{\lceil (t-t_0)\bar{a}M/a_{max}(M+N)\rceil} + p_2\left(1 - \xi^{\lceil (t-t_0)\bar{a}M/a_{max}(M+N)\rceil}\right), & t_0 < t \le n_a, \\
p_{f1}\left\{\xi^{\lceil (t-t_0)\bar{a}M/a_{max}(M+N)\rceil} + p_2\left(1 - \xi^{\lceil (t-t_0)\bar{a}M/a_{max}(M+N)\rceil}\right)\right\} \\
+(1-p_{f1})\xi^{\lceil (t-n_a)\bar{a}M/a_{max}(M+N)\rceil(1-p_2)}\mathbf{E}\left\{r^i_{m_h}(n_a)\right\}, & t \ge n_a.
\end{cases} \\
&\mathbf{E}\left\{\left[r^i_{m_h}(t+1)\right]^2\right\} = \\
&\begin{cases}
1, & t \le t_0, \\
\xi^{2\lceil (t-t_0)\bar{a}M/a_{max}(M+N)\rceil} + p_2\left(1 - \xi^{2\lceil (t-t_0)\bar{a}M/a_{max}(M+N)\rceil}\right), & t_0 < t \le n_a, \\
p_{f1}\left\{\xi^{2\lceil (t-t_0)\bar{a}M/a_{max}(M+N)\rceil} + p_2\left(1 - \xi^{2\lceil (t-t_0)\bar{a}M/a_{max}(M+N)\rceil}\right)\right\} \\
+(1-p_{f1})\xi^{2\lceil (t-n_a)\bar{a}M/a_{max}(M+N)\rceil(1-p_2)}\mathbf{E}\left\{r^i_{m_h}(n_a)\right\}^2, & t \ge n_a.
\end{cases}
\end{aligned}
\tag{4.27}
$$

4.3.3.2 Expectation of Rational Nodes' Reputation

Then we analyze the expectation of reputation for rational nodes. Similarly, the correlation of current reputation and the reputation of first time slot could be written as

$$
r^i_{r_h}(t+1) = \xi^t r^i_{r_h}(1) + (1-\xi)\sum_{j=2}^{t+1} \xi^{t+1-j}\hat{r}^i_{r_h}(j). \tag{4.28}
$$

In the same way, $\mathbf{E}\left\{r^i_{r_h}(t+1)\right\}$, $\mathbf{E}\left\{\left[r^i_{r_h}(t+1)\right]^2\right\}$ and $\mathbf{E}\left\{\left[r^i_{r_h}(t+1)\right]^3\right\}$ are derived as

$$\begin{aligned}
\mathbf{E}\left\{r^i_{r_h}(t+1)\right\} &= \xi^t r^i_{r_h}(1) + \left(1-\xi^t\right)\mathbf{E}\left\{\hat{r}^i_{r_h}(j)\right\} \\
\mathbf{E}\left\{\left[r^i_{r_h}(t+1)\right]^2\right\} &= \xi^{2t}\left[r^i_{r_h}(1)\right]^2 + (1-\xi^t)^2\mathbf{E}\left\{\left[\hat{r}^i_{r_h}(j)\right]^2\right\} \\
&\quad +2r^i_{r_h}(1)\xi^t(1-\xi^t)\mathbf{E}\left\{\hat{r}^i_{r_h}(j)\right\} \\
\mathbf{E}\left\{\left[r^i_{r_h}(t+1)\right]^3\right\} &= \xi^{3t}\left[r^i_{r_h}(1)\right]^3 + (1-\xi^t)^3\mathbf{E}\left\{\left[\hat{r}^i_{r_h}(j)\right]^3\right\} \\
&\quad +3r^i_{r_h}(1)\xi^t(1-\xi^t)^2\mathbf{E}\left\{\left[\hat{r}^i_{r_h}(j)\right]^2\right\} \\
&\quad +3\left[r^i_{r_h}(1)\right]^2\xi^{2t}(1-\xi^t)\mathbf{E}\left\{\hat{r}^i_{r_h}(j)\right\}. \qquad (4.29)
\end{aligned}$$

For a rational node, he/she averagely has $\bar{k} = \sum_{k=0}^{\infty}\lambda(k)k$ neighbors. Thus, a rational user's reputation value could be regarded to update every $(M+N)/\bar{k}$ time slots. Since rational users' evaluation of reputation doesn't involve the differences in strategies, there only exists two cases: $t \le t_0$ and $t > t_0$. When $t > t_0$, the only factor affecting rational users' reputation is the authenticity of information forwarded t_0 time slots ago, so the probability of $\hat{r}^i_{m_h}(t)$ being 0 is $p'_f(1-p_1)$, where p'_f is the population state t_0 time slots ago. Then the expectation of $\hat{r}^i_{m_h}(t)$, $\left[\hat{r}^i_{m_h}(t)\right]^2$ and $\left[\hat{r}^i_{m_h}(t)\right]^3$ after t_0 is $(p'_f p_1 + 1 - p'_f)$ rather than p_1. Similarly, the expectation of reputation for rational nodes is shown as

$$\begin{aligned}
&\mathbf{E}\left\{r^i_{r_h}(t+1)\right\} = \\
&\begin{cases} 1, & t \le t_0, \\ \xi^{\lceil (t-t_0)\bar{k}/(M+N)\rceil} + (p'_f p_1 + 1 - p'_f)\left(1-\xi^{\lceil (t-t_0)\bar{k}/(M+N)\rceil}\right), & t \ge t_0. \end{cases} \\
&\mathbf{E}\left\{\left[r^i_{r_h}(t+1)\right]^2\right\} = \\
&\begin{cases} 1, & t \le t_0, \\ \xi^{2\lceil (t-t_0)\bar{k}/(M+N)\rceil} + (p'_f p_1 + 1 - p'_f)\left(1-\xi^{2\lceil (t-t_0)\bar{k}/(M+N)\rceil}\right), & t \ge t_0. \end{cases} \\
&\mathbf{E}\left\{\left[r^i_{r_h}(t+1)\right]^3\right\} = \\
&\begin{cases} 1, & t \le t_0, \\ \xi^{3\lceil (t-t_0)\bar{k}/(M+N)\rceil} + (p'_f p_1 + 1 - p'_f)\left(1-\xi^{3\lceil (t-t_0)\bar{k}/(M+N)\rceil}\right), & t \ge t_0. \end{cases} \qquad (4.30)
\end{aligned}$$

Substituting (4.27) and (4.30) into (4.17) and (4.22), the final form of population dynamics of two types nodes could be derived, based on which we are able to predict every small change in the process, get detailed evolutionary dynamics and finally foretell corresponding ESSs by setting dynamics to zero.

4.4 Simulations and Experiments

In this section, we use synthetic and real-world data to evaluate the smart evolution model. First of all, the theoretical analysis of dynamics and ESSs is validated over synthetic social networks with malicious nodes. Then, the effectiveness of the model is further verified based on Facebook networks and the data set of microblog information dissemination.

4.4.1 Synthetic Networks

In the simulation of synthetic networks, we synthesize a uniform network with the constant degree $k = 25$ to simulate the information diffusion process. For the network, we generate totally $a_{max} = 10$ malicious users, $M = 500$ type I rational users and $N = 1000$ type II rational users. The malicious neighbors of type I nodes are assumed to be evenly distributed, i.e., $\mu(a) = 1/10,\ 1 \leq a \leq 10$. Selection intensity α is set as 0.04, which satisfies the requirement of weak selection. Other parameters including influence coefficient ξ, probability of rational users sending true information p_1, probability of malicious users sending true information p_2, time threshold t_0 and n_a are set as 0.5, 0.9, 0.6, 1500 and 100000 respectively.

We first assess the analysis for population state by comparing the smart evolution scheme with the scheme without reputation mechanism, where the payoff matrix is set as $PM1 : u_{ff} = 0.3, u_{fn} = 0.8, u_{nn} = 0.2$. Results of the smart evolution scheme are shown in Fig. 4.3a, and we could see that theoretic analysis matches well with the simulation results. Due to direct connections between malicious nodes and type I nodes, p_{f1} is higher than p_{f2} and p_f. From the figure, we could also see that there is a big drop after the time slot $t = n_a = 100000$. This is because when n, the duration of malicious users' strategy $s = 0$, exceeds the time that type I node could endure, the reputation of malicious users would decline significantly. In this way, the impact of malicious users will be smaller than before, and the population state would inevitably decrease accordingly. Figure 4.3b shows the results of the scheme without reputation mechanism, which means that reputation values of all users are 1 and users treat every neighbor equally. It could be observed that without the reputation mechanism, there is no drop and the population states are higher than those in Fig. 4.3a, demonstrating the effectiveness of the reputation mechanism. With the smart evolution model, the ESS of whole network decreases by 10.13% before time slot n_a and 12.62% after time slot n_a, compared with the scheme without reputation mechanism. Therefore, the smart evolution model could effectively reduce the negative effects of malicious users.

Next, we implement simulations with different payoff matrices. Figure 4.4 shows the results under the payoff matrix $PM2 : u_{ff} = 0.8, u_{fn} = 0.3, u_{nn} = 0.2$. The setting for payoffs is corresponding to the situation that forwarding information is beneficial to all users. Hence, the reputation mechanism makes no difference and

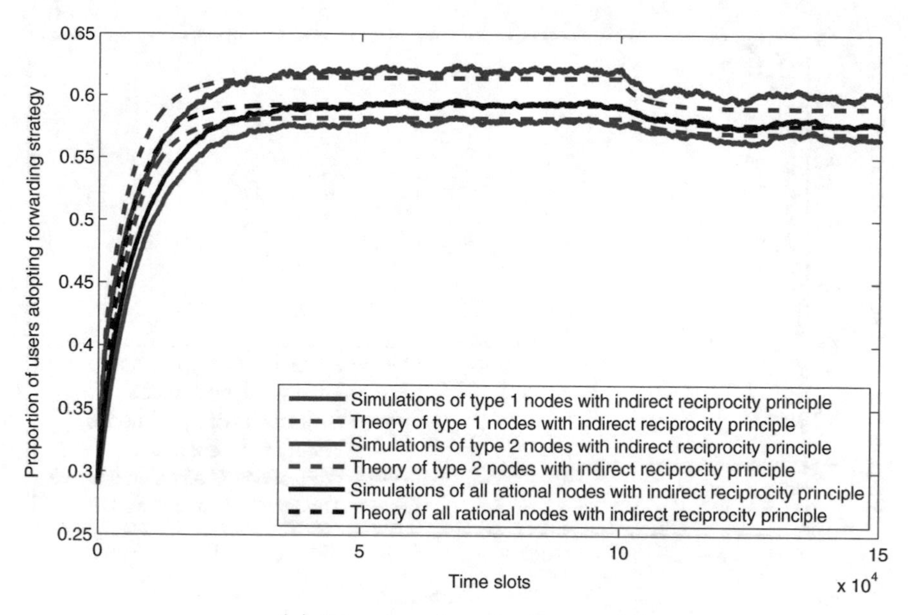

(a) The smart evolution model.

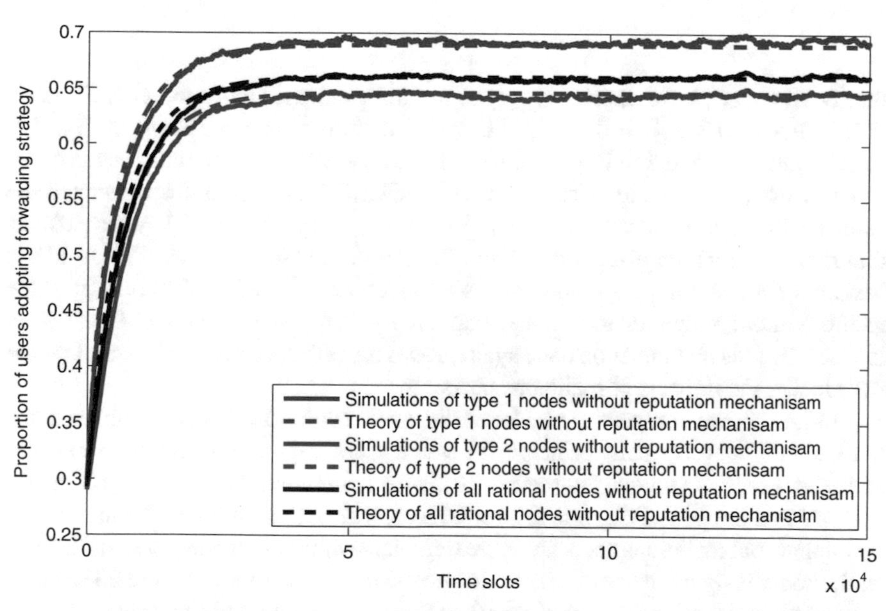

(b) Scheme without reputation mechanism.

Fig. 4.3 Simulation results of population states for two schemes under the payoff matrix PM1: $u_{ff} = 0.3$, $u_{fn} = 0.8$, $u_{nn} = 0.2$

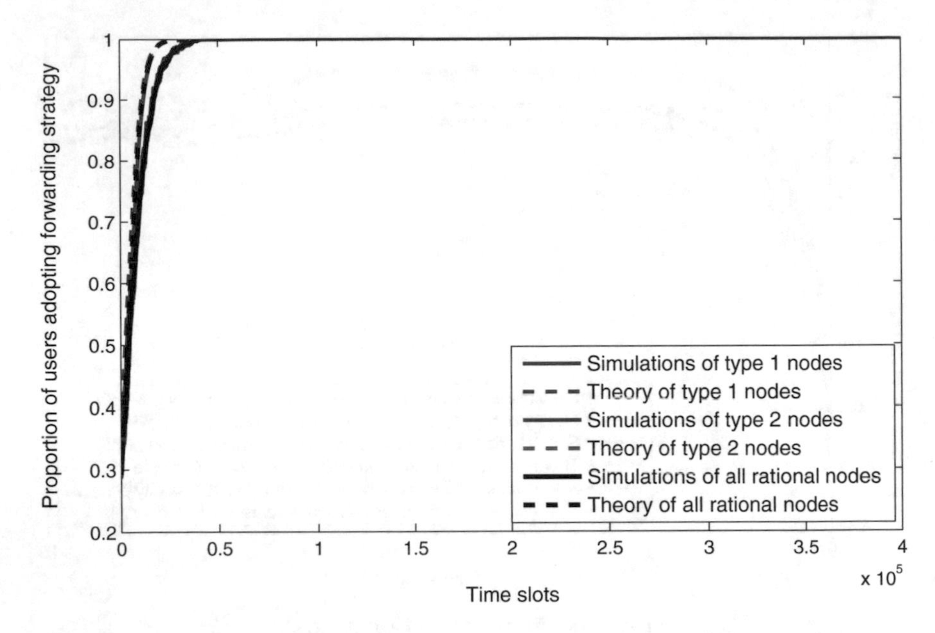

Fig. 4.4 Simulation results of population states under the payoff matrix PM2: $u_{ff} = 0.8$, $u_{fn} = 0.3$, $u_{nn} = 0.2$

finally all users would choose to spread the information. Under the payoff matrix $PM3 : u_{ff} = 0.3, u_{fn} = 0.8, u_{nn} = 0.4$, population states are shown in Fig. 4.5, which is similar to that in Fig. 4.3a. It could be observed that population state drops more at time slot n_a compared with that under $PM1$. Figure 4.6 illustrates the results with payoff matrix $PM4 : u_{ff} = 0.1, u_{fn} = 0.1, u_{nn} = 0.3$, which means that rational users are discouraged from forwarding the information. We could see that after time slot n_a, population states all reach zero, which is different from the results without reputation mechanism that p_f, p_{f1} and p_{f2} would not be 0, as shown in Chap. 3. This phenomenon once again proves the effectiveness of the smart model in reducing the impacts of malicious users.

In Fig. 4.7, we compare the ESSs of the smart evolution model and the scheme without indirect reciprocity principle, in which users' reputation wouldn't be shared and every user has a unique judgement for his/her neighbors. We set the payoff matrix as $PM1$ and the p_1 as 0.9. p_2 is set from 0 to 1, with 0.1 as interval. The increase of p_2 implies that malicious users are more reliable, resulting in the increase of ESSs as well. From the figure, it could be seen that no matter $t < n_a$ or $t > n_a$, the ESSs of the scheme without reciprocity are higher than those in the model of this chapter. This is because with indirect reciprocityprinciple, users share reputation to help each other identify their new neighbors reliability, due to which users can know their neighbors more thoroughly. Notice that when $p_2 = 0$, all ESSs are 0.5494 and in such a case malicious users could be ignored. It could also be seen that as p_2 increases, the drop of ESSs from $t < n_a$ to $t > n_a$ becomes bigger, which indicates that when malicious

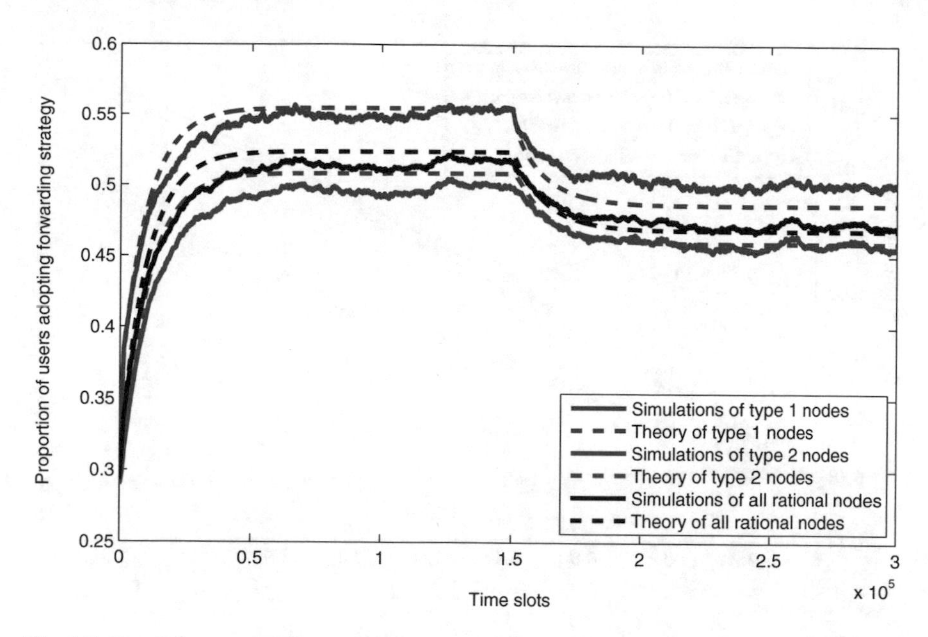

Fig. 4.5 Simulation results of population states under the payoff matrix PM3: $u_{ff} = 0.3, u_{fn} = 0.8, u_{nn} = 0.4$

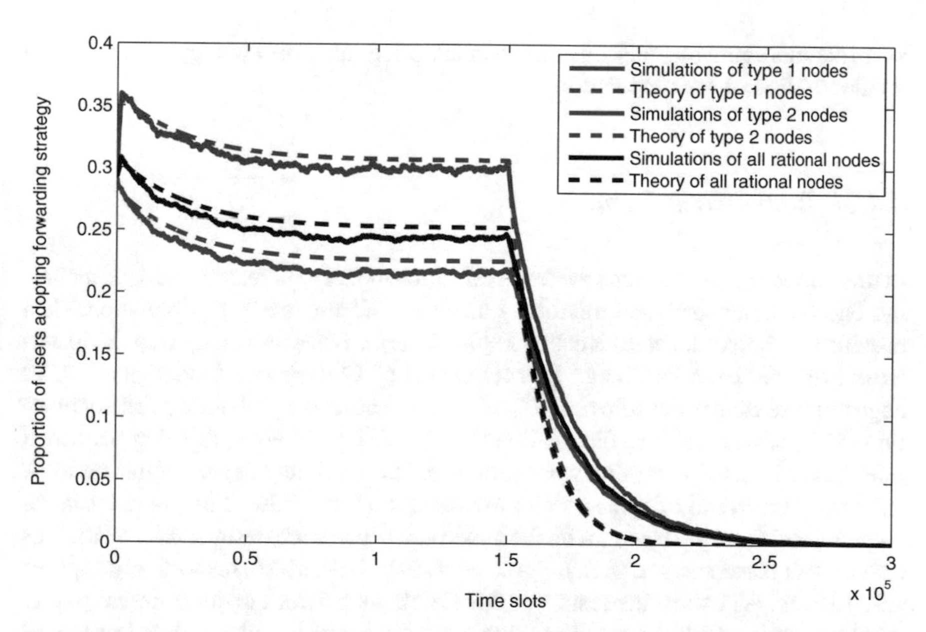

Fig. 4.6 Simulation results of population states under the payoff matrix PM4: $u_{ff} = 0.1, u_{fn} = 0.1, u_{nn} = 0.3$

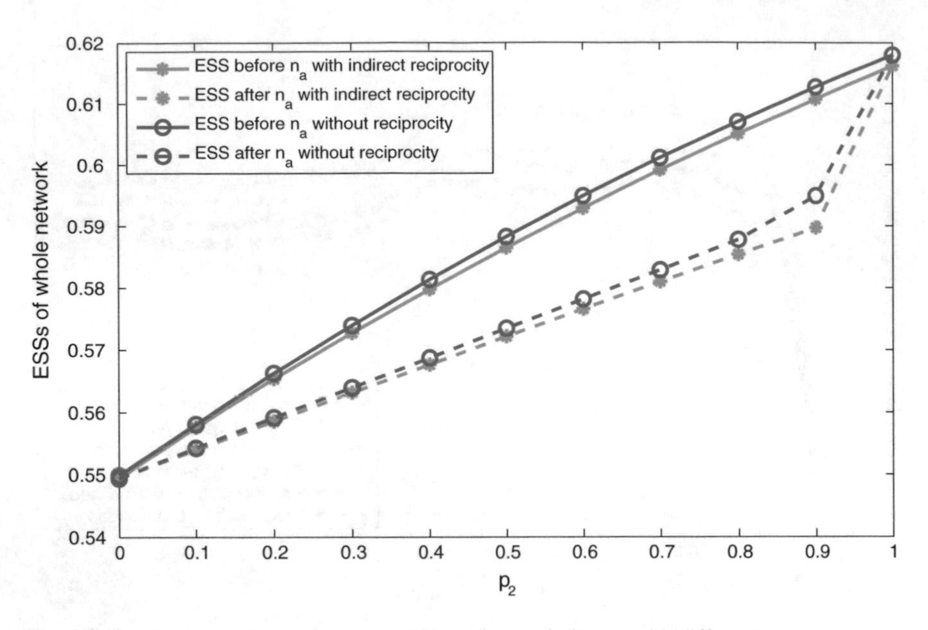

Fig. 4.7 Comparison of two schemes on ESSs before and after n_a with different p_2

users are more reliable, their insistence in adopting the same strategy would lead to greater decline of their reputation.

4.4.2 Real-World Data

In this subsection, we evaluate the smart evolution model with real-world data, including Facebook networks and microblog data sets. We first perform simulations on a specific Facebook subnetwork called *socfb-Haverford76*, a university friendship network extracted from Facebook. It is composed of 1446 people (nodes) and 59589 edges representing friendship ties [32]. Figure 4.8 shows the probability density function (PDF) and cumulative distribution function (CDF) of network's degree, which reflect the characteristics of network structure. The maximum degree of this network is 375, and the average degree $\bar{k} = 82$. We assume a total of 20 malicious users in the network, and the top 20 nodes with the maximum degrees are designated as malicious users. Other parameters α, p_1 and p_2 are set as 0.05, 0.9 and 0.6, respectively. Figures 4.9, 4.10 and 4.11 show the results of this Facebook network under different payoff matrices, from which we could see that theoretic analysis and simulated results fit well. Comparing the results of the smart evolution model and the scheme without reputation mechanism, we could find that in every moment the population states of the smart evolution scheme are larger, especially when u_{nn} is greater than u_{ff}. It demonstrates the effectiveness of the model in real networks. Compared with the

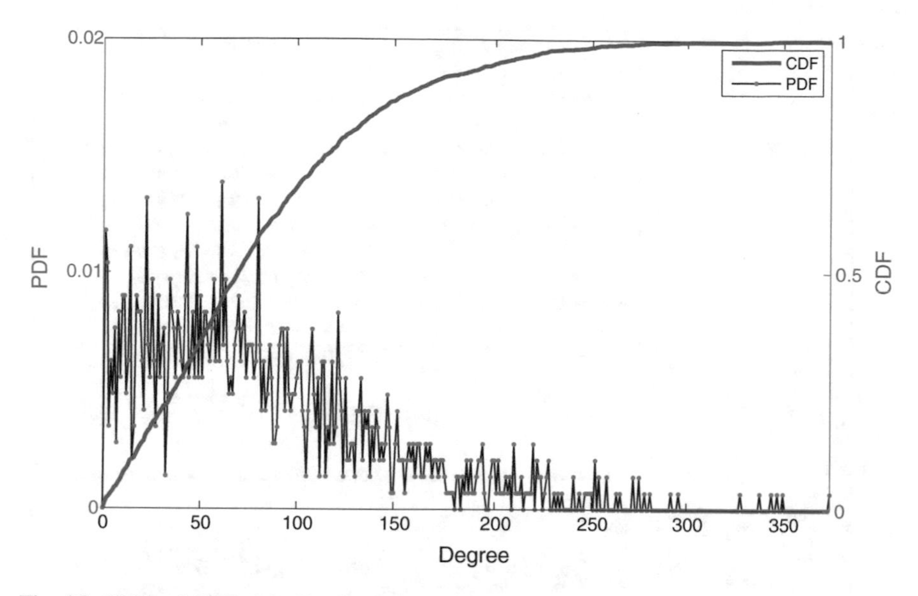

Fig. 4.8 PDF and CDF of the Facebook network *socfb-Haverford76*

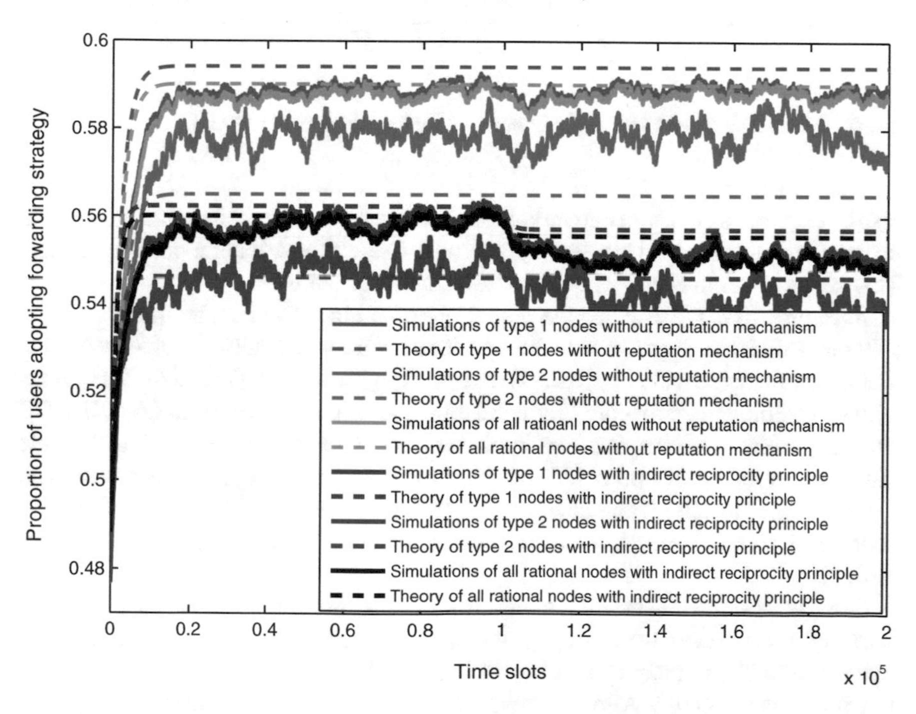

Fig. 4.9 Simulation results for *socfb-Haverford76* with payoff matrix PM1

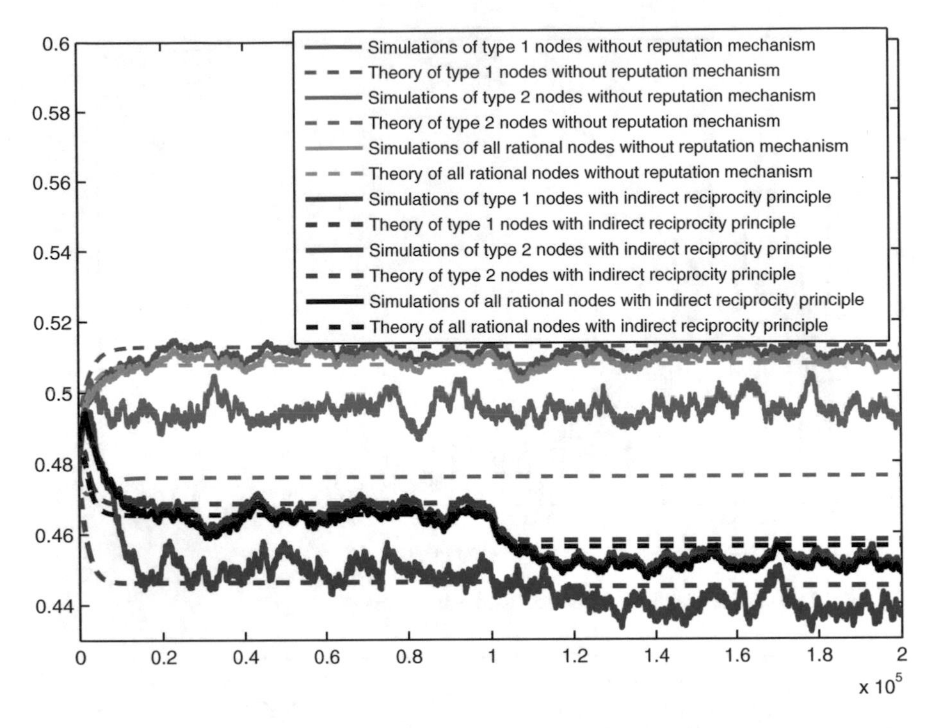

Fig. 4.10 Simulation results for *socfb-Haverford76* with payoff matrix PM3

results of previous uniform network, there is a big fluctuation of simulated diffusion process and some gaps between the simulated results and theoretical predictions. This is due to the uneven and widespread distribution of users.

In Figure 4.12, we evaluate the model over the other five Facebook networks in [32] under different payoff matrix settings. The theoretical results, denoted by dashed lines, are calculated from (4.1), (4.17) and (4.22) by setting (4.17) and (4.22) to zero, while the simulations, denoted by solid lines, are obtained by simulating the DB strategy update rule over five Facebook networks and then getting the final stable states. From the figure, we could see that the simulation analysis agrees well with the theoretical results. It can also be seen that when the payoff matrix is $PM4$, or in other words $u_{nn} > \max\{u_{ff}, u_{fn}\}$, the decline of ESSs after n_a time slots is more distinct.

Next, we test the model using microblog information spreading data set, which is extracted from Weibo, the most popular Chinese microblogging service [33, 34]. It contains 300,000 popular microblog diffusion episodes, including publishing time and processed user ID. After a new piece of information is released, users who are also interested in will forward it, and this process can be seen as information diffusion. To make the experiment matched with the scene of model, we select four episodes with negative contents, and the number of users involved in these episodes is 19950. Among these users, we sort out those who repost several episodes in the

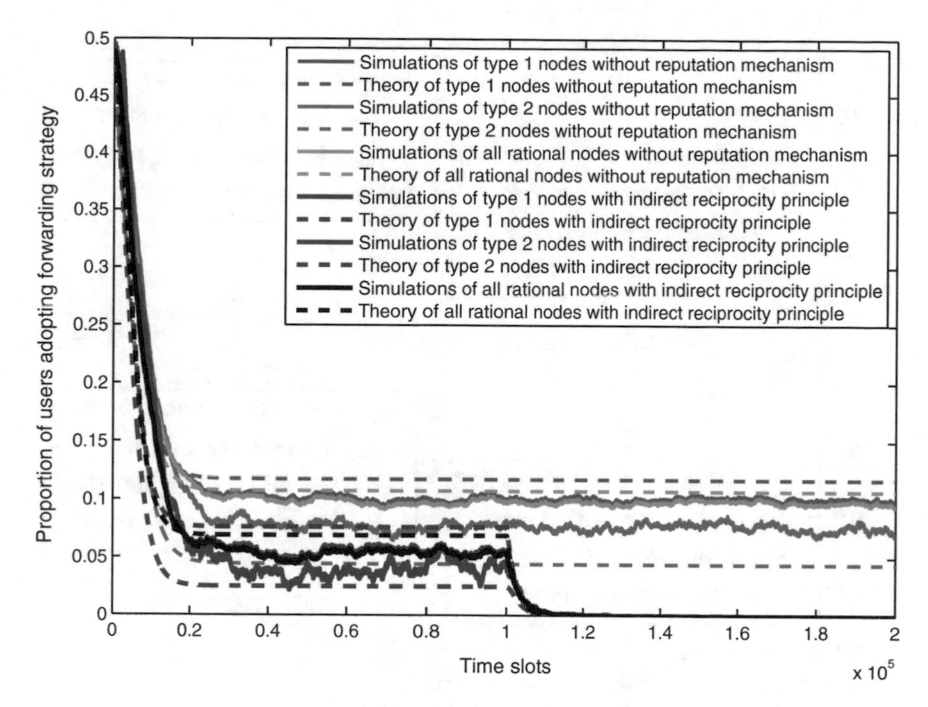

Fig. 4.11 Simulation results for *socfb-Haverford76* with payoff matrix PM4

whole data set as type I users, and the rest who only repost once are type II nodes, based on which we could get $M = 775$ and $N = 19175$. Then, the ESS of one piece of information in the diffusion is calculated as the proportion of users retweeting this information among all users. Since the network structure is unknown in the data set, we assume that every user has same degree $k = 100$, which is common in real social network. Actually the network degree does not influence the results much owing to the large scale of network.

Different from previous experiments that set parameters at first and then conduct simulations, in this simulation we first estimate parameters based on the data set, then test the smart evolution model by checking ESSs. According to (4.1), (4.17) and (4.22), we estimate four parameters α, Φ, Φ_n, a_{max} which are related to information feature and network characteristics, by least square method using the Matlab function *lsqcurvefit*. The estimated parameters of four groups are presented in Table 4.1. The results of α do not seem to meet the assumption of the smart evolution model. In fact, α is related to payoffs and the time length of diffusion process, and it would multiply with all these factors. From the table, we could also see that all groups satisfy $\Phi + \Phi_n = u_{ff} - u_{fn} < 0$ and $\Phi_n = u_{fn} - u_{nn} > 0$, implying that u_{fn} is the biggest. That is, forwarding the information is only beneficial for some people.

Then we conduct simulations based on the estimated parameters. Figure 4.13 shows the results of four groups, comparing the actual ESSs from the data set, i.e., the results without reputation scheme, and the ESS using the smart evolution scheme

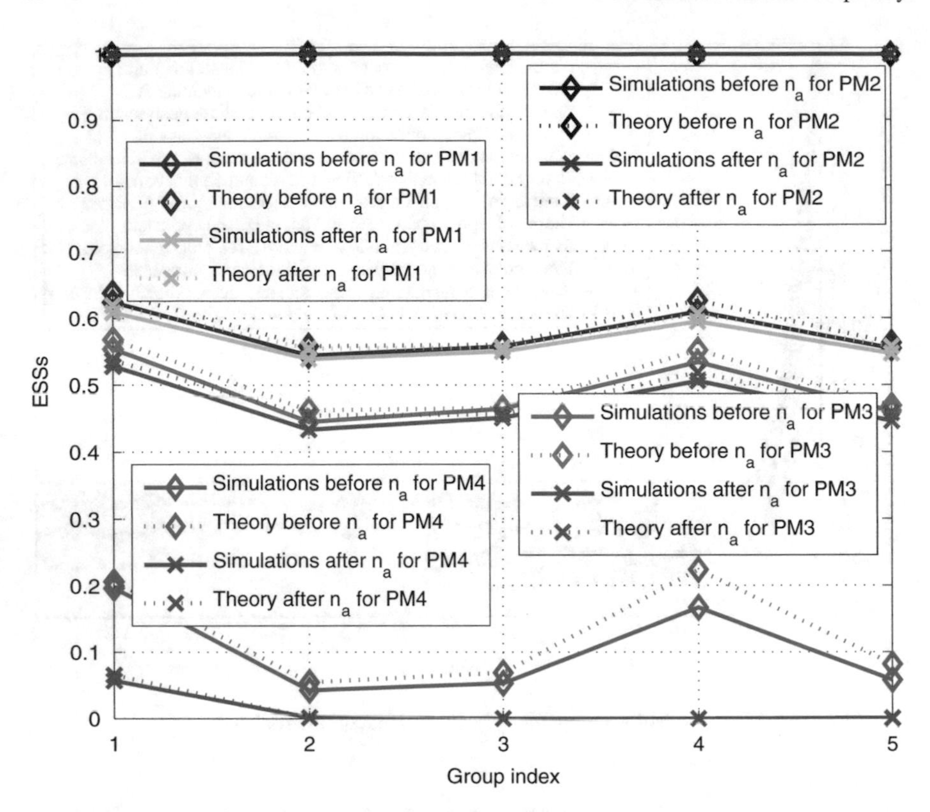

Fig. 4.12 ESSs of whole network over five Facebook networks with different payoff matrices

Table 4.1 Estimated Parameters

Group index	Selection strength (α)	Payoffs combination 1 (Φ)	Payoffs combination 2 (Φ_n)	Number of malicious nodes (a_{max})
1	8.0523	−10.1588	5.7156	10
2	12.3930	−12.1119	2.7657	10
3	9.0545	−10.3482	6.7007	13
4	15.9471	−15.6106	3.3765	11

with reputation scheme, where p_1 and p_2 are set to 0.9 and 0.6 respectively. With the reputation mechanism in the smart evolution model, there is a significant drop of ESSs compared with original results. Also, we could see that the theoretic ESSs match well with the simulated ones. In Fig. 4.14, we reveal the effects of p_1 and p_2 on simulated ESSs over the uniform network in Group 1. Obviously the change of p_1 has a greater impact on the final ESSs, while the change of p_2 has an almost zero effect on ESSs, the reason of which is that compared with the number of type II

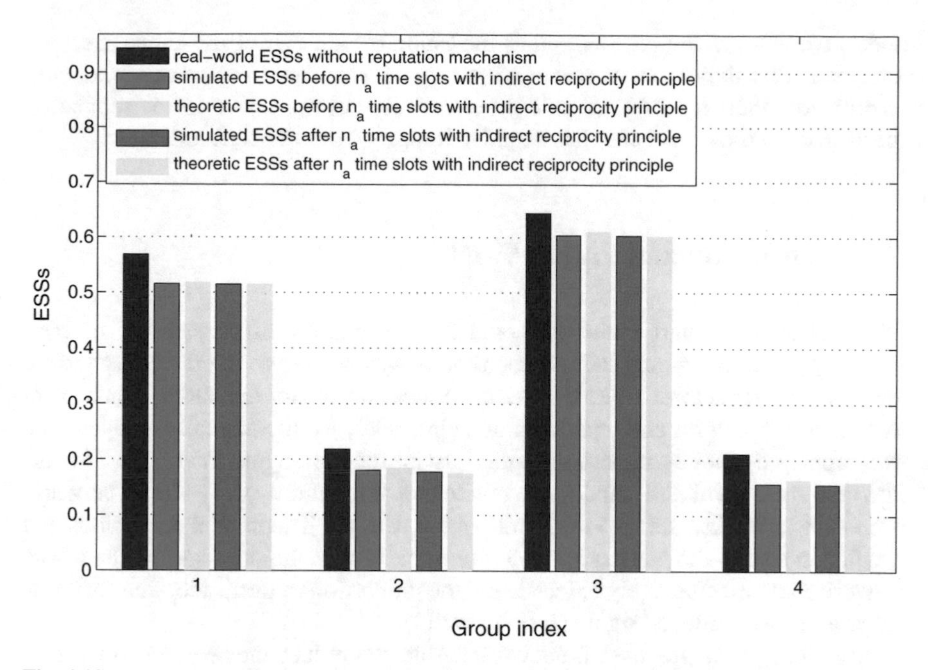

Fig. 4.13 ESSs of actual results and the smart evolution scheme with parameters estimated from microblogging network

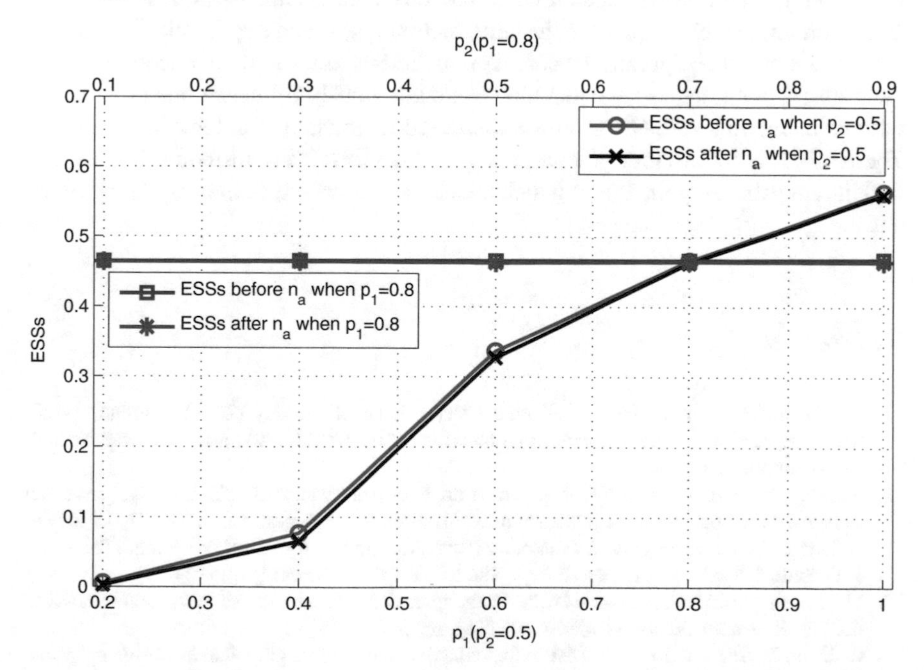

Fig. 4.14 Simulated ESSs of group 1 with different p_1 and p_2

nodes, type I users' number as well as the estimated number of malicious users are too small. This shows that no matter which type of users are inclined to spread false information, their reputation would be effectively cut down due to the reputation mechanism in the model and thus negative impacts would be reduced.

4.5 Conclusion and Future Work

In this chapter, a smart evolution model for information diffusion based on evolutionary game theory and indirect reciprocity was proposed, to effectively reduce the negative impacts of malicious users. According to the reputation mechanism including social norm and reputation updating rules, we theoretically analyzed the evolutionary dynamics, expectation of all users' reputation and final ESSs. To validate the smart evolution model, we conducted simulations on synthetic network, Facebook networks, and real-world microblog data set. Theoretical derivations and simulation results showed that the negative impacts of malicious users on the whole network could be effectively reduced, and the proportion of users adopting forwarding strategy was reduced significantly as well.

Even though the proposed framework has helped reduce the negative impacts of malicious users, there are still lots of issues to be explored. For example, we still need to find more effective and efficient methods to detect and deter malicious users and to minimize their impacts on the entire network as much as possbile. In addition, given the indirect reciporcity mechanism, malicious users will also need to change their attacking strategies such that they can continuously influence others' opinions without being attacked. Given so, we also need to update the defense mechanisms. Therefore, it is important to theoretically analyze this "cat-and-mouse" game, and find the equilibrium points from which attackers and system designers will not deviate.

References

1. N. Martin, How much data is collected every minute of the day (2019). [Online]. Available: https://www.forbes.com/sites/nicolemartin1/2019/08/07/how-much-data-is-collected-every-minute-of-the-day
2. X. Deng, Y. Dou, T. Lv, Q.V.H. Nguyen, A novel centrality cascading based edge parameter evaluation method for robust influence maximization. IEEE Access **5**, 22119–22131 (2017)
3. S. Tadelis, *Game Theory: An Introduction* (Princeton University Press, Princeton, 2013)
4. J. Weibull, *Evolutionary Game Theory* (The M.I.T. Press, Cambridge, 1995)
5. I.I. Hussein, An individual-based evolutionary dynamics model for networked social behaviors,' in *2009 American Control Conference* (2009), pp. 5789–5796
6. C. Jiang, Y. Chen, K.J.R. Liu, Distributed adaptive networks: a graphical evolutionary game-theoretic view. IEEE Trans. Signal Process. **61**(22), 5675–5688 (2013)
7. C. Jiang, Y. Chen, K.J.R. Liu, Graphical evolutionary game for information diffusion over social networks. IEEE J. Sel. Top. Signal Process. **8**(4), 524–536 (2014)

8. C. Jiang, Y. Chen, K.J.R. Liu, Evolutionary dynamics of information diffusion over social networks. IEEE Trans. Signal Process. **62**(17), 4573–4586 (2014)
9. X. Cao, Y. Chen, C. Jiang, K.J. Ray Liu, Evolutionary information diffusion over heterogeneous social networks. IEEE Trans. Signal Inf. Process. Over Netw. **2**(4), 595–610 (2016)
10. Y. Li, B. Qiu, Y. Chen, H.V. Zhao, Analysis of information diffusion with irrational users: A graphical evolutionary game approach, in *ICASSP 2019—2019 IEEE International Conference on Acoustics, Speech and Signal Processing (ICASSP)* (2019), pp. 2527–2531
11. DOMO, Data never sleeps 7.0 (2019). [Online]. Available: https://www.domo.com/learn/data-never-sleeps-7
12. G. Tong, W. Wu, S. Tang, D. Du, Adaptive influence maximization in dynamic social nctworks. IEEE/ACM Trans. Netw. **25**(1), 112–125 (2017)
13. Z. Wang, Y. Yang, J. Pei, L. Chu, E. Chen, Activity maximization by effective information diffusion in social networks. IEEE Trans. Knowl. Data Eng. **29**(11), 2374–2387 (2017)
14. S. Agarwal, S. Mehta, Social influence maximization using genetic algorithm with dynamic probabilities, in *2018 Eleventh International Conference on Contemporary Computing (IC3)* (2018), pp. 1–6
15. S.T. Hasson, E. Akeel, Influence maximization problem approach to model social networks, in *2019 International Conference on Advanced Science and Engineering (ICOASE)* (2019), pp. 135–140
16. G. Song, X. Zhou, Y. Wang, K. Xie, Influence maximization on large-scale mobile social network: a divide-and-conquer method. IEEE Trans. Parallel Distrib. Syst. **26**(5), 1379–1392 (2015)
17. H. Wu, J. Shang, S. Zhou, Y. Feng, B. Qiang, W. Xie, Laim: A linear time iterative approach for efficient influence maximization in large-scale networks. IEEE Access **6**, 44221–44234 (2018)
18. H. Zhang, D.T. Nguyen, H. Zhang, M.T. Thai, Least cost influence maximization across multiple social networks. IEEE/ACM Trans. Netw. **24**(2), 929–939 (2016)
19. H. Nguyen, R. Zheng, On budgeted influence maximization in social networks. IEEE J. Sel. Areas Commun. **31**(6), 1084–1094 (2013)
20. D. Goldenberg, A. Sela, E. Shmueli, Timing matters: Influence maximization in social networks through scheduled seeding. IEEE Trans. Comput. Soc. Syst. **5**(3), 621–638 (2018)
21. S. Dhamal, P.K.J., Y. Narahari, Information diffusion in social networks in two phases. IEEE Trans. Netw. Sci. Eng. **3**(4), 197–210 (2016)
22. N. Arazkhani, M.R. Meybodi, A. Rezvanian, An efficient algorithm for influence blocking maximization based on community detection, in *2019 5th International Conference on Web Research (ICWR)* (2019), pp. 258–263
23. J. Zheng, L. Pan, Least cost rumor community blocking optimization in social networks, in *2018 Third International Conference on Security of Smart Cities, Industrial Control System and Communications (SSIC)* (2018), pp. 1–5
24. L. Fan, Z. Lu, W. Wu, B. Thuraisingham, H. Ma, Y. Bi, Least cost rumor blocking in social networks, in *2013 IEEE 33rd International Conference on Distributed Computing Systems* (2013), pp. 540–549
25. B. Wang, G. Chen, L. Fu, L. Song, X. Wang, Drimux: dynamic rumor influence minimization with user experience in social networks. IEEE Trans. Knowl. Data Eng. **29**(10), 2168–2181 (2017)
26. M.A. Nowak, K. Sigmund, Evolution of indirect reciprocity. Nature **437**(7063), 1291 (2005)
27. H. Ohtsuki, Y. Iwasa, M.A. Nowak, Indirect reciprocity provides only a narrow margin of efficiency for costly punishment. Nature **457**(7225), 79 (2009)
28. Y. Chen, K.R. Liu, Indirect reciprocity game modelling for cooperation stimulation in cognitive networks. IEEE Trans. Commun. **59**(1), 159–168 (2010)
29. L. Xiao, Y. Chen, W.S. Lin, K.R. Liu, Indirect reciprocity security game for large-scale wireless networks. IEEE Trans. Inf. Forensics Secur. **7**(4), 1368–1380 (2012)
30. B. Zhang, Y. Chen, J.-L. Yu, B. Chen, Z. Han, Indirect-reciprocity data fusion game and application to cooperative spectrum sensing. IEEE Trans. Wirel. Commun. **16**(10), 6571–6585 (2017)

31. H. Ohtsukia, M.A. Nowak, The replicator equation on graphs. J. Theor. Biol. **243**, 86–97 (2006)
32. R.A. Rossi, N.K. Ahmed, The network data repository with interactive graph analytics and visualization, in *AAAI* (2015). [Online]. Available: http://networkrepository.com
33. J. Zhang, B. Liu, J. Tang, T. Chen, J. Li, Social influence locality for modeling retweeting behaviors, in *Proceedings of the 23rd International Joint Conference on Artificial Intelligence (IJCAI'13)*, pp. 2761–2767
34. J. Zhang, J. Tang, J. Li, Y. Liu, C. Xing, Who influenced you? predicting retweet via social influence locality. ACM Trans. Knowl. Discov. Data **9**(3) (2015). [Online]. Available: https://doi.org/10.1145/2700398

Chapter 5
Diffusion of Multi-source Correlated Information

Abstract Recently, online social networks are playing an ever-important role in both our social life and economy. Therefore, modeling of information diffusion over social networks is a common research topic, which is of crucial importance to better understand how the avalanche of information overflow leads to the detrimental consequences, and how to motivate some beneficial information spreading. However, most model-based works on information diffusion either merely consider the spreading of one single message or make the assumption that different diffusion processes are independent of each other. In real-world scenarios, multi-source correlated information often spread together, which jointly influence users' decisions. In this chapter, we aim to model the multi-source information diffusion processes from a graphical evolutionary game perspective. Specifically, we model users' local interactions and strategic decision making process, and analyze the evolutionary dynamics of the diffusion processes of correlated information. We conduct simulations on both synthetic and Facebook real-world networks, and the simulation results are consistent with our theoretical analysis. We also test the model on the users' forwarding data in "Weibo" social networks and observe an effective prediction performance on the real-world information spreading processes.

Keywords Correlated information diffusion · Evolutionary game theory · Population dynamics · Relationship dynamics · Influence dynamics

5.1 Introduction

In recent years, with the popularity of smartphones and the advance of networking technologies, online social networks have rapidly emerged, where information spreads in a sharing, editable and interactive pattern. Unlike the previous scenarios that people only receive news from the websites, nowadays every user has the ability to create, modify and spread the information in the online social networks. However, it is an established fact that the convenience of spreading information can also be misused in a malicious manner. For instance, gossips and rumors propagate across online social platforms, and sometimes result in information overflow and severe

Y. Chen and H. V. Zhao, *Behavior and Evolutionary Dynamics in Crowd Networks*, Lecture Notes in Social Networks, https://doi.org/10.1007/978-981-15-7160-2_5

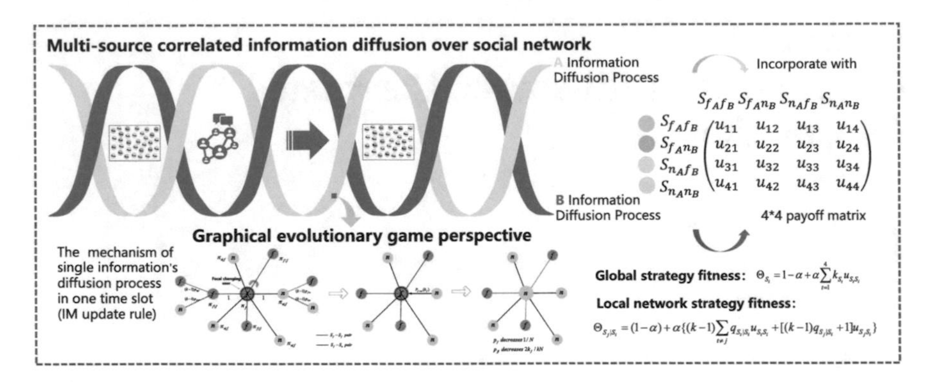

Fig. 5.1 Modeling mulit-source correlated information diffusion from graphical evolutionary game perspective

damage to the economy. Therefore, many researchers aimed to design models to understand the mechanisms of the information diffusion process (Fig. 5.1).

There have been many related works studying the information diffusion process in online social networks [1–3], which mainly can be classified into two categories: empirical data-driven approaches and theoretical model-based analysis. Specifically, the data-driven approaches usually utilize the machine learning algorithms to learn the parameters from large amounts of data. For instance, some statistical feature extraction mechanisms were extensively elaborated in [4–8]. In [4], the authors aimed to develop a data-driven framework that models the "forward probability" of a user's adoption of information and statistically learns how different contagions interact with each other in the Twitter dataset. However, users' interactions were ignored, and the learned "cluster" concept lacked corresponding real-world meanings. In [5], extensive mechanisms were designed to address these issues. Nevertheless, their approach did not have a large improvement in prediction performance but has a high sensitivity and strict requirement of the training dataset. Li et al. [6] studied limitations of previous data driven based works and developed a statistical framework from the micro-perspective. They jointly utilized the global and local influence of users to calculate time-related social payoffs based on real data (Sina Weibo and Flickr). However, their approach did not consider diffusion of correlated information.

In contrast to the data-driven approaches, most model-based approaches tended to establish the mathematical dynamic models to analyze the diffusion system's dynamic process. Inspired by different disciplines such as psychology, sociology and economics, these models usually have strong interpretability and extensibility. Some novel models include the epidemic model [9], independent cascade (IC) model [10], linear threshold (LT) model [11] and their variantions [12–15]. These models help investigate and understand many meaningful issues, e.g., influence maximization, accurate personalized marketing and breaking news detection. However, users' interactions and imitation processes were usually ignored in these models and analyses.

The information diffusion process in online social networks usually can be described as follows: when a piece of information is released by someone, his or her neighbors (friends) may be influenced and decide whether to forward the information or not. Therefore, both the users' personal interests in the information and the influence of their friends may determine the diffusion process of the information. The interactions among users and the users' decision-making processes should be taken into consideration in the model.

The works in [16, 17] provided an evolutionary game-theoretic approach to study different information diffusion scenarios. Considering that all data are generated by users in social networks, it is important to jointly consider users' interactions [18] and the statistical characteristics of whole diffusion's data. Evolutionary game theory (EGT) [19] was originally used to model the biological evolution. Therefore, it comes naturally that we can take users' information exchange, interaction and forwarding process as an analogy of the complex natural selection process, aiming to investigate the underlying principles in high-dynamic information diffusion processes. This promising framework can jointly model users' interactions at the micro level and study the dynamic information diffusion process, suggesting a suitable and tractable paradigm to address the challenge of analyzing the complex information diffusion process in online social networks [20].

This framework has been applied to many areas of signal processing and been extended to more special scenarios. In [20], the authors attempted to utilize the framework in heterogeneous networks (each user has a diverse preference, thus, different payoff matrix). The work in [21] took irrational users (those who always intentionally forward fake news) into consideration. And in [22–24], the framework was extended to image processing and communication networking areas, suggesting the extensibility of this promising tool.

In this chapter, unlike the previous graphical EGT information diffusion models that consider the spreading of one single message and assume different diffusion processes are independent of each other, we aim to study the multi-source correlated information diffusion processes, which is more consistent with the real-world scenarios. By modeling users' decision making as an evolutionary game, we analyze three diffusion dynamics (population dynamics, relationship dynamics and influence dynamics) specifically. To verify the model's correctness and effectiveness, we conduct experiments not only on the synthetic networks but also on real-world Facebook networks. Simulation results are consistent with our theoretical analysis.

We also conduct experiments to learn the model's parameters from the crawled Sina Weibo users' forwarding data at correlated hashtags. With only 18 changeable parameters, the model's prediction results can also fit well with the real-world diffusion data. At the same time, we limit the proportion of the training dataset and observe that 60% of the crawled data can help fit the whole information-spreading process with less than 8% error rate, showing the potentials of our model in prediction of the correlated information diffusion process.

The rest of this chapter is organized as follows. In Sect. 5.2, we formulate the graphical EGT framework for multi-source information diffusion scenarios. In Sect. 5.3, we rederive three evolutionary dynamics in the 4*4 payoff matrix to extend the previous graphical EGT framework for correlated information diffusion. The Monte-Carlo simulations are shown in Sect. 5.4. The conclusions are drawn in Sect. 5.5.

5.2 Problem Fomulation of the Multi-source Information Diffusion

In this section, we will fist introduce the basic concepts of graphical evolutionary game theory briefly, and then formulate the multi-source information diffusion process over social networks.

Specifically, in social network, graph structure refers to the topology of the social network where the information is propagating. The nodes represent users and edges represent social relationships between users, e.g., friendship in Facebook. As for each user, he or she may have several strategies in the game, i.e forward the information or not.

The utility function in the evolutionary game is usually defined as "fitness". In the information diffusion scenario, users with larger fitness values tend to have higher influence on others. Specifically, the fitness is calculated by $\pi = (1 - \alpha)B + \alpha U$ where B is the baseline fitness indicating users' standing in the network and U is the fitness getting from interactions with others. The parameter α represents the selection intensity, i.e, the relative contribution of the game to the fitness value. In this chapter, without loss of generality, we let $B = 1$ and consider the simple scenario where a user's baseline fitness has a larger impact on fitness than users' interactions, thus we have $\alpha \to 0$.

However, in the previous graphical EGT information diffusion models that consider only one single message and assume different diffusion processes are independent of each other. The strategies of users are forward the message or not. In this chapter, we consider a simple multi-source information diffusion scenario, where two correlated messages A and B are spreading simultaneously in the network. Therefore, users' strategy set $\left\{S_{f_A f_B}, S_{f_A n_B}, S_{n_A f_B}, S_{n_A n_B}\right\}$ can be defined as follows:

$$\begin{cases} S_{f_A f_B}, & \text{forward both message A and B,} \\ S_{f_A n_B}, & \text{forward message A but not B,} \\ S_{n_A f_B}, & \text{forward message B but not A,} \\ S_{n_A n_B}, & \text{forward neither message A nor B.} \end{cases} \tag{5.1}$$

In the following part, we abbreviate the strategy set as $\{S_1, S_2, S_3, S_4\}$.

Furthermore, we consider the symmetric scenario where when a user with strategy S_i meets another user adopting strategy S_j, both receive the same payoff $u_{s_i s_j}$. Therefore, the payoff matrix is defined as $\left\{u_{s_i s_j}\right\}_{4\times 4}$ where $u_{s_i s_j}$ denotes the payoff of a user with strategy S_j meeting a user with strategy S_i, regardless of other users' influence. In the network, the percentages of users adopting the four strategies are denoted as $\left\{k_{S_1}, k_{S_2}, k_{S_3}, k_{S_4}\right\}$, and let the average fitness of each strat-

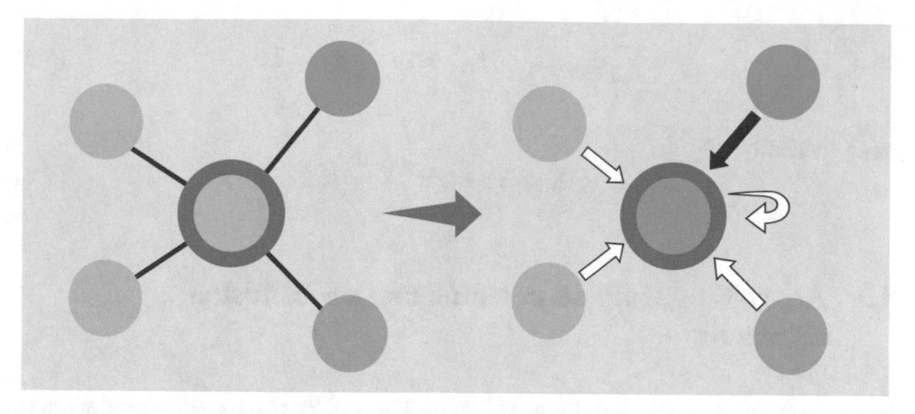

Fig. 5.2 The IM update rule. The focal selected user will change or remain his/her strategy according to each strategy's fitness in his/her local neighbor network

egy be $\{\Theta_{S_1}, \Theta_{S_2}, \Theta_{S_3}, \Theta_{S_4}\}$. Therefore, the average fitness of adopting strategy S_i can be calculated as

$$\Theta_{S_i}(k_{S_1}, k_{S_2}, k_{S_3}, k) = (1-\alpha)B + \alpha \sum_{t=1}^{4} k_{S_t} u_{s_t s_i}. \tag{5.2}$$

As the information propagates in the social network, users may change their adopted strategies and imitate their friends' strategies to achieve higher fitness. Specifically, the strategy evolution process is divided into numerous time slots. In each time slot, one user is selected randomly as the focal user to update the strategy with a certain update rule. In this chapter, we adopt the imitation update rule (IM), which is originated from the biology evolving theory. The analysis of other strategy update rules, such as the DB and BD rules, is similar and omitted here. As shown in Fig. 5.2, in the IM update rule, the selected focal user will either imitate the strategy of one neighbor or keep his/her current strategy unchanged. The probability of imitating the neighbor's strategy (including his/her current strategy) is proportional to each strategy's fitness in the selected focal user's local neighborhood.

However, this strategy updating process happens in the local neighborhood of the selected focal user rather than the entire network, we define four kinds of local networks according to the focal user's adopted strategy: Q_1, Q_2, Q_3, Q_4, where $Q_i = \{q_{s_1|s_i}, q_{s_2|s_i}, q_{s_3|s_i}, q_{s_4|s_i}\}$ models the local neighborhood where the focal user adopts strategy S_i. Here, $q_{s_j|s_i}$ denotes the percentage of users adopting strategy S_j in the local network where the selected focal user's adopted strategy is S_i. Therefore, the whole ***local network state*** can be denoted as $Q = \{Q_1^T, Q_2^T, Q_3^T, Q_4^T\} = \{q_{s_j|s_i}\}_{4\times 4}$. Let $X = (p_{s_1}, p_{s_2}, p_{s_3}, p_{s_4})^T$ denote the ***global network state***, where p_{s_i} is the percentage of users adopting strategy S_i in the entire network. Define $\Xi = (1, 1, 1, 1)^T$, and the local network state and the global network state satisfy the following relationship:

$$\sum_{i=1}^{4} q_{s_i|s_j} p_{s_i} = p_{s_j}, \text{ s.t. } \sum_{i} p_{s_i} = 1, \tag{5.3}$$

or equivalently,

$$QX = X, \text{ s.t. } \Xi^T X = 1. \tag{5.4}$$

5.3 Analysis of Multi-source Information Diffusion Dynamics

In this section, with the definitions of the global and local network states, we will analyze three dynamics of multi-source information diffusion over social networks:

- population dynamics $\dot{p}_{s_i}$, $(i = 1, 2, 3, 4)$ that model how users in the entire network change their strategies from time to time;
- relationship dynamics $\dot{p}_{s_j s_i}$, $(1 \leq i, j \leq 4)$ that model the dynamics of the edges connecting different nodes in the entire network; and
- influence dynamics $\dot{q}_{s_j|s_i}$, $(1 \leq i, j \leq 4)$ that model how users influence each other in a local network.

This investigation helps us better understand how users influence each other's decision process both locally and globally, and enables derivation of the final evolutionarily stable states (ESS).

5.3.1 Analysis of the Population Dynamics

The population dynamics are the dynamics of the percentages of users in the entire network adopting different strategies. In the following, we will analyze different scenarios, aiming to derive the closed-form expression of the population dynamics.

5.3.1.1 Scenario 1: $\Delta p_{S_i} = -\frac{1}{N}$

We first study the scenario that a user previously adopting strategy S_i changes his or her strategy to other strategies. Specifically, according to the IM update rule, the randomly selected focal user is with strategy S_i. Then, he or she will imitate the $S_j (j \neq i)$ strategy, proportional to the fitness of the corresponding strategy. In this scenario, the percentage of strategy S_i will decrease by one unit ($\Delta p_{S_i} = -\frac{1}{N}$).

We assume that the selected user has k neighbors, among which k_{S_1}, k_{S_2}, k_{S_3}, k_{S_4} users adopt strategy S_1, S_2, S_3, S_4, respectively. Since this is a multinomial distribution, the probability of such a scenario is

$$\Psi^{de}_{(S_i;k_{S_1},k_{S_2},k_{S_3},k_{S_4})} = \frac{k!}{k_{S_1}!k_{S_2}!k_{S_3}!k_{S_4}!} q_{s_1|s_i}^{k_{S_1}} \cdot q_{s_2|s_i}^{k_{S_2}} \cdot q_{s_3|s_i}^{k_{S_3}} \cdot q_{s_4|s_i}^{k_{S_4}}. \tag{5.5}$$

Among the focal user's k nerghbors, for the neighbors adopting $S_j (j \neq i)$, each of them has $(k-1)q_{s_t|s_j}$ neighbors using strategy $S_t (t \neq i)$, as well as $1 + (k-1)q_{s_i|s_j}$ neighbors using strategy S_i on average, where "1" refers to the selected focal user with strategy S_i. Therefore, the average fitness of the focal user's neighbors who adopt strategy S_j can be calculated using

$$\Theta_{s_j|s_i} = (1-\alpha) + \alpha\{(k-1)\sum_{t\neq j} q_{s_t|s_i} u_{s_t s_i} + [(k-1)q_{s_j|s_i} + 1]u_{s_j s_i}\}. \tag{5.6}$$

According to the IM update rule, the probability of a user updating his/her strategy is proportional to the fitness of each strategy in the local network. Therefore, the probability that the focal user updates his/her current strategy from S_i to $S_j (j \neq i)$ is

$$\mathbb{E}\left[P_{S_i \to S_j}(k_{S_1}, k_{S_2}, k_{S_3}, k)\right] = \frac{k_{S_j}\Theta_{S_j|S_i}}{\sum_t k_{S_t}\Theta_{S_t|S_i} + \Theta_{S_i}}. \tag{5.7}$$

The probability that the focal user updates his/her current strategy from S_i to any other strategies can be summarized as

$$\mathbb{E}\left[P_{S_i \to others}\right] = 1 - \frac{\Theta_{(S_i;k_{S_1},k_{S_2},k_{S_3},k)} + k_{S_i}\Theta_{S_i|S_i}}{\Theta_{(S_i;k_{S_1},k_{S_2},k_{S_3},k)} + \sum_t k_{S_t}\Theta_{S_t|S_i}}. \tag{5.8}$$

Therefore, the probability of $\Delta p_{S_i} = -\frac{1}{N}$ can be calculated as

$$\text{Prob}\left(\Delta p_{S_i} = -\tfrac{1}{N}\right) = \sum_{k_{S_3}=0}^{k-k_{S_1}-k_{S_2}} \sum_{k_{S_2}=0}^{k-k_{S_1}} \sum_{k_{S_1}=0}^{k} p_{S_i} \cdot \Psi^{de}_{S_i} \cdot \mathbb{E}\left[P_{S_i \to others}\right]. \tag{5.9}$$

5.3.1.2 Scenario 2: $\Delta p_{S_i} = \frac{1}{N}$

Similarly, we can analyze the reverse scenario that the selected focal user previously adopting strategy $S_j (j \neq i)$ changes to strategy S_i. Therefore, the percentage of users adopting strategy S_i will increase by one unit ($\Delta p_{S_i} = \frac{1}{N}$).

The average fitness of the focal S_j user's neribors who adopt S_i strategy $\Theta_{s_i|s_j}$ can also be derived using the same method as above. Accordingly, the probability that the focal user changes his/her strategy from $S_j (j \neq i)$ to S_i can be written as

$$\mathbb{E}\left[P_{S_j \to S_i}(k_{S_1}, k_{S_2}, k_{S_3}, k)\right] = \frac{k_{S_i}\Theta_{S_i|S_j}}{\sum_t k_{S_t}\Theta_{S_t|S_j} + \Theta_{S_j}}. \tag{5.10}$$

Accordingly, the probability that the focal user updates his/her current strategy from any other strategy to strategy S_i can be summarized as:

$$\mathbb{E}\left[P_{other\rightarrow S_i}\right]=\frac{k_{S_i}\Theta_{S_i|S_j}}{\Theta_{(S_j;k_{S_1},k_{S_2},k_{S_3},k)}+\sum_t k_{S_t}\Theta_{S_t|S_j}}. \tag{5.11}$$

In the configuration of $\Psi^{in}_{(S_j;k_{S_1},k_{S_2},k_{S_3},k_{S_4})}$, the probability of $\Delta p_{S_i}=\frac{1}{N}$ can be calculated as

$$\text{Prob}\left(\Delta p_{S_i}=\tfrac{1}{N}\right)=\sum_{j,j\neq i}\sum_{k_{S_3}=0}^{k-k_{S_1}-k_{S_2}}\sum_{k_{S_2}=0}^{k-k_{S_1}}\sum_{k_{S_1}=0}^{k}p_{s_j}\cdot\Psi^{in}_{S_j}\cdot\mathbb{E}\left[P_{others\rightarrow S_i}\right]. \tag{5.12}$$

Combining the above derivation, we have the probability of p_{s_i} increases by $\frac{1}{N}$ and decreases by $\frac{1}{N}$, respectively. Therefore, the expected variation of p_{s_i} can be calcultaed per unit time, which is the definition of the population dynamic p_{s_i}

$$\dot{p}_{S_i}=\frac{1}{N}\text{Prob}\left(\Delta p_{S_i}=\frac{1}{N}\right)-\frac{1}{N}\text{Prob}\left(\Delta p_{S_i}=-\frac{1}{N}\right). \tag{5.13}$$

Note from the above analysis, in order to study the population dynamics, we need to first analyze the relationship dynamics and the influence dynamics. For example, in the above, the unknown variable in $\dot{p}_{s_i}$ is the local network state $q_{s_i|s_j}$, which is the topic of the following discussion.

5.3.2 *Analysis of Relationship Dynamics*

The relationship dynamics $\dot{p}_{s_is_j}$, $(1\leq i,\ j\leq 4)$ are the dynamics of the global edge states, modeling the dynamics of relationship among users. Note that the relationship dynamics satisfy the constraint

$$\sum_{1\leq i,j\leq 4}\dot{p}_{s_is_j}=0. \tag{5.14}$$

Here, $p_{s_is_j}$ denotes the percentage of the edge connecting a user using strategy S_i with a user adopting strategy S_j. As the states of both nodes may change, to avoid repeated calculation, in the following, when analyzing $\dot{p}_{s_js_i}$, we only consider the scenario where one node changes. Without loss of generality, we study the scenario where the node previously adopting strategy S_i changes his/her strategy.

Note that

$$q_{s_i|s_j}p_{s_j}=q_{s_j|s_i}p_{s_i}\text{ and }p_{s_is_j}=p_{s_js_i}. \tag{5.15}$$

Therefore, there are two possibilities, when $i \neq j$ and when $i = j$.

- Scenario 1: $i \neq j$

When $i \neq j$, there are also two sub-scenarios. The first corresponds to the scenario where the randomly selected user previously adopts strategy S_j and then imitates a neighbor using strategy $S_l(l \neq j)$, which causes the edge state $p_{s_i s_j}$ to decrease by one unit. This probability can be calculated as follows:

$$P_{s_i s_j}^{D_A} = p_{s_j} \cdot \sum_{l \neq j} q_{s_l|s_j}(k-1)q_{s_i|s_l}. \tag{5.16}$$

The other case is where the selected user previously uses strategy $S_l(l \neq j)$, then copies a neighbor adopting strategy S_j. Thus, the edge state $p_{s_i s_j}$ increases by one unit, and this probability is:

$$P_{s_i s_j}^{I_A} = \sum_{l \neq j} p_{s_l} q_{s_j|s_l}(k-1)q_{s_i|s_j} + p_{s_i} q_{s_j|s_i}. \tag{5.17}$$

Summarizing (5.16) and (5.17), with the help of above analysis and relations in (5.15), the relationship dynamics can be calculated as:

$$\dot{p}_{s_i s_j}^{A} = \frac{NP_{s_i s_j}^{I_A} - NP_{s_i s_j}^{D_A}}{Nk/2} = \frac{2}{k}\left[(k-1)\, p_{s_j} \sum_{l} q_{i|l} q_{l|j} - k q_{i|j}\right]. \tag{5.18}$$

- Scenario 2: $i = j$

When we choose a user who previously chooses strategy S_i user, then copies a neighbor adopting strategy $S_l(l \neq i)$, the edge state $p_{s_i s_j}$ decreases by one unit, and this probability is

$$P_{s_i s_j}^{D_B} = p_{s_i} \sum_{l \neq i} q_{s_l|s_i}\left[(k-1)q_{s_i|s_l} + 1\right]. \tag{5.19}$$

In the reverse case, the focal user first uses strategy $S_l(l \neq i)$, then changes to strategy S_i. Accordingly, the edge state $p_{s_i s_j}$ increases by one unit, whose probability is

$$P_{s_i s_j}^{I_B} = p_{s_i} q_{s_i|s_i}(k-1) \sum_{l \neq i} q_{s_l|s_i}. \tag{5.20}$$

Same as the analysis in (5.18), the relationship dynamic of this scenario can be calculated as:

$$\dot{p}_{s_i s_j}^{B} = \frac{NP_{s_i s_j}^{I_B} - NP_{s_i s_j}^{D_B}}{Nk/2} = \frac{2}{k}\left[(k-1)\, p_{s_j} \sum_{l} q_{i|l} q_{l|j} - k q_{i|j} + 1\right]. \tag{5.21}$$

As we can see, the relationship dynamics only have one unkown variable, local network state $q_{s_i|s_j}$, which leads us to the following analysis of the influence dynamics.

5.3.3 *Analysis of the Influence Dynamics*

The influence dynamics model the dynamics in a local network and quantifies the influence of a user on his/her neighbors. For instance, $\dot{q}_{s_i|s_i} \rightarrow 1$ means most neighbors of the focal user tend to choose the same strategy as the focal user, suggesting the focal user may be a public figure and has high impact on others.

Considering the weak selection scenario with $\alpha \rightarrow 0$, the local network status changes at a faster speed than the global network states, and we can assume that the local neighborhood states converge much faster than the global dynamics. Therefore, we can regard the global network state as constant and derive the dynamics of the local network as follows:

$$\dot{q}_{i|j} = \frac{\dot{p}_{s_is_j}}{\dot{p}_{s_j}} = \frac{\dot{p}_{s_is_j}}{p_{s_j}} + o(\alpha) = \frac{2}{k}\left[(k-1)\,p_i \sum_l q_{s_i|s_l} q_{s_l|s_i} - k q_{s_i|s_i} + \delta_{ij}\right] + o(\alpha). \tag{5.22}$$

Given the above analysis of the influence dynamics, when $\dot{q}_{s_i|s_j} = 0$, we can obtaion the temporarily steady local network state $q^*_{s_i|s_j}$, which is the only unkown variable $q_{s_i|s_j}$ in *Population Dynamics* and *Relationship Dynamics*.

According to the $\dot{q}_{s_i|s_j} = 0$ condition, we can have the following:

$$I + (k-1)\,Q^2 - kQ = 0. \tag{5.23}$$

where I is the identity matrix. When we take $QX = X$ and $\Xi^T X = 1$ into consideration, $(X\Xi^T)X = X$, $XI = I$. Local state $Q(i.e., \{q_{s_i|s_j}\}_{4\times4})$ can be written as a linear function of $X\Xi^T$ and X. Therefore, we can obtain the unique non-trivial solution:

$$Q = \frac{k-2}{k-1} X\Xi^T + \frac{1}{k-1} I. \tag{5.24}$$

Hence, after substituting the temporarily steady local network state $q_{s_i|s_j} = \frac{(k-2)p_{s_i} + \delta_{ij}}{k-1}$ into the population dynamic expression and consider the weak selection ($\alpha \ll 1$), we can simplify the closed-form solution of population dynamics as follows:

$$\dot{p}_{s_i} = \alpha' p_{s_i} \cdot \sum_j p_{s_j}\left[\Delta_{ij} - (k^2 + k - 6)\sum_i p_{s_i} u_{ji}\right] = 0, \tag{5.25}$$

where

$$\begin{cases} \Delta_{ij} = (k+3)u_{ii} + (k^2 - k - 2)u_{ij} - 3u_{ji} - (k+3)u_{jj}, \\ \alpha' = \alpha \frac{k(k-2)}{(k+1)^2(k-1)}. \end{cases} \tag{5.26}$$

Then we can numerically solve the above equations and find the evolutionarily steady state $\{p^*_{s_i}\}$.

5.4 Experiments

In this section, several experiments are conducted to verify the correctness of model's theoretical derivation, the information diffusion dynamics specifically. We also attempt to demonstrate the model's effectiveness of real-world's applications.

5.4.1 Synthetic Networks and Real-World Networks

First, we conduct simulations to verify correctness of model's derivation. The designed simulation framework can be illustrated as follows (Fig. 5.3):

In our studied scenarios, the information diffusion is a process with large randomness but has statistical features at the same time. Accordingly, Monte-Carlo algorithm that relies on numerous repeated random sampling is applicable to investigate the underlying principles.

The experiments are conducted by simulating the fitness calculation and strategy update processes step-by-step. The network structure is repeatedly re-generated for 20 times and the simulation experiments are conducted in the same network for 32 times to reduce randomness. In each run, the fitness calculation and strategy update processes have 400 iterations, ensuring the experiment has a sufficient time span. The network size is set to 2000 users. The intensity of selection α is set as 0.01, i.e., weak selection. The initial percentage of each strategy is set as $\{0.1, 0.2, 0.3, 0.4\}$.

We extensively develop the experiments in the random regular network, Erdös Rényi (ER) random networks and Barabási-Albert (BA) scale-free networks (the specific methods are referred to [17]) to prevent biased results based on one particular realization of a specific network type. Moreover, we compare the model-based curves and the simulation results in these three synthetic networks, as we can see in Figs. 5.4 and 5.5. In Fig. 5.6, the gap for Barabási-Albert (BA) scale-free network is due to the neglected dependence between the global network state and the network degree, which is analyzed in [17] specifically.

In the experiment of the real-world network structure, we choose the Facebook dataset in [25], which contains totally 4039 users and 88234 relations, and the average degree approximately equals 40.We depicted the graph structure of Facebook

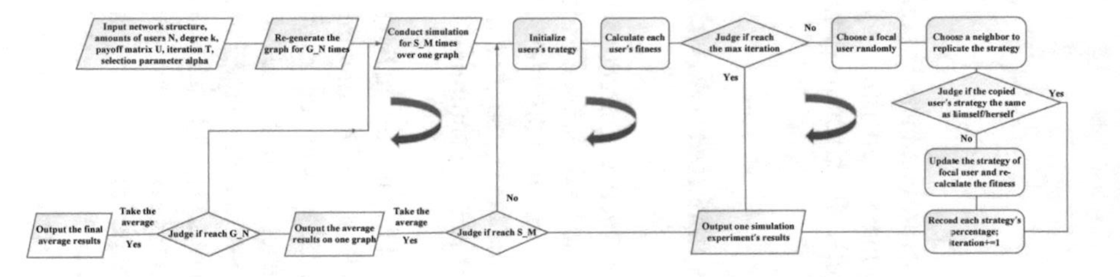

Fig. 5.3 The framework of the designed simulation expriments, in the aim of verifing models's theoretical derivation, the information diffusion dynamics specifically

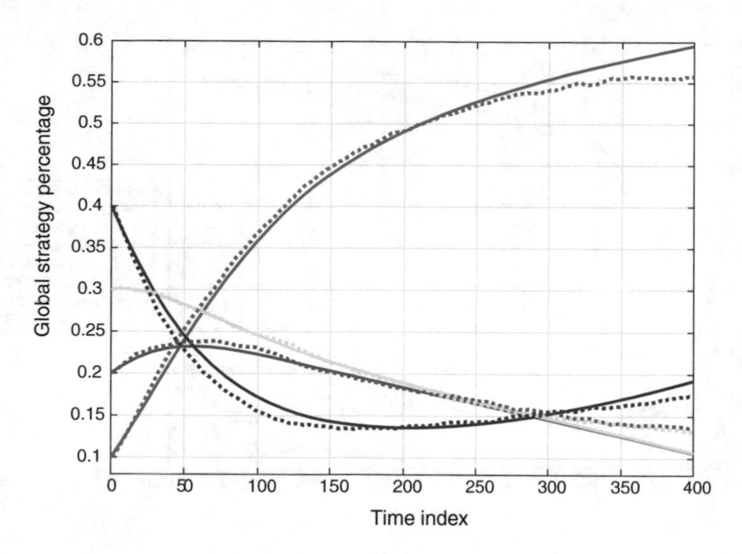

Fig. 5.4 Simulation results of the random regular network

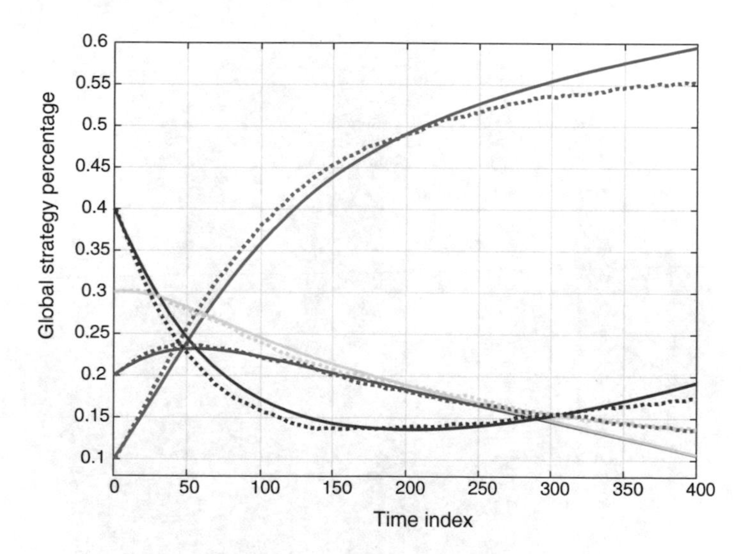

Fig. 5.5 Simulation results of the ER random network

network in Fig. 5.7, where the nodes represent users and the edges represent relationship.

It can be seen in Fig. 5.7 that the real-world Facebook socail network has many clusters and communities of strong combination. The degree of each user is quite different in the network as well, which is in accordance with the real-world scenarios.

Then, we apply our theoretical models in this real-world network structure. The simulation results and theoretical prediction shows the experiment results are ploted

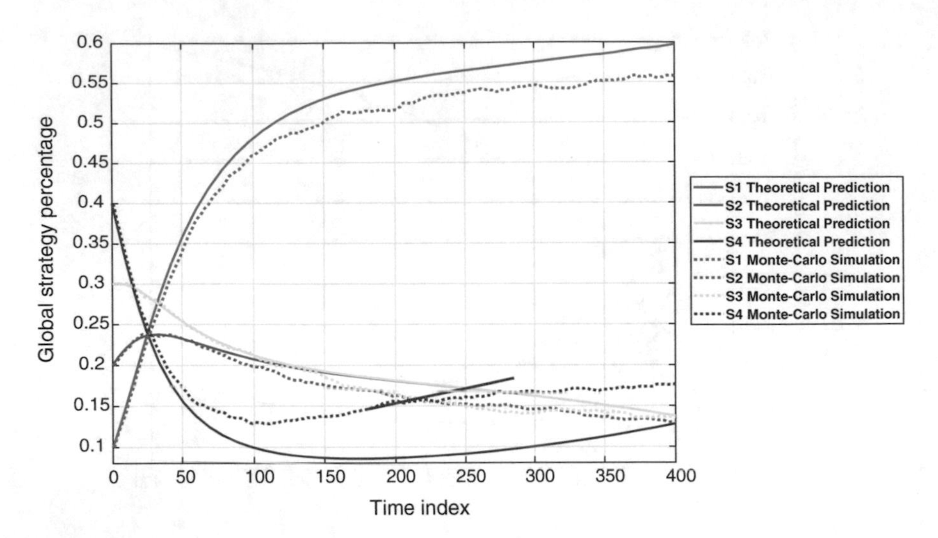

Fig. 5.6 Simulation results of the BA scale free network

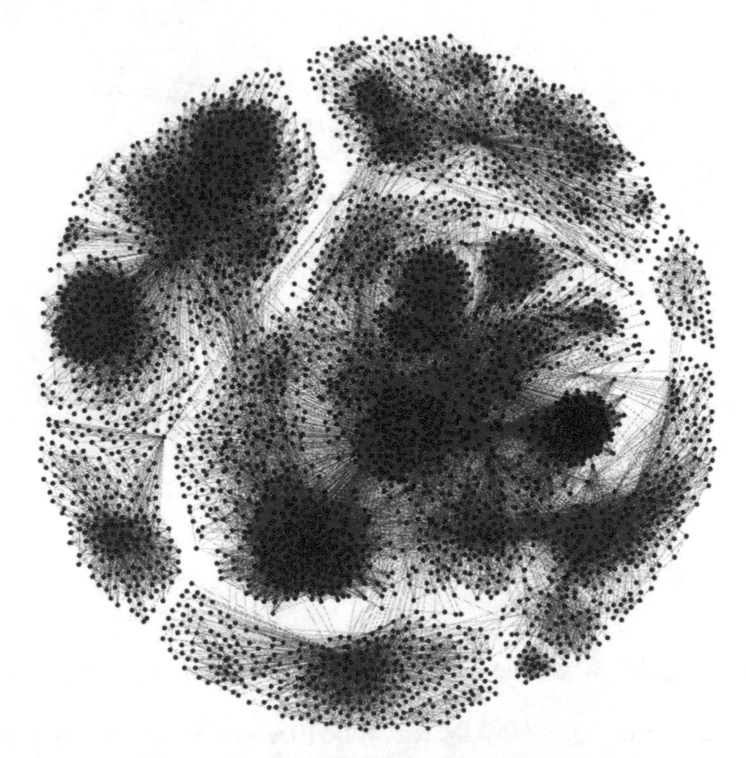

Fig. 5.7 The illustrated graph structure of the facebook real-world social network for experiment

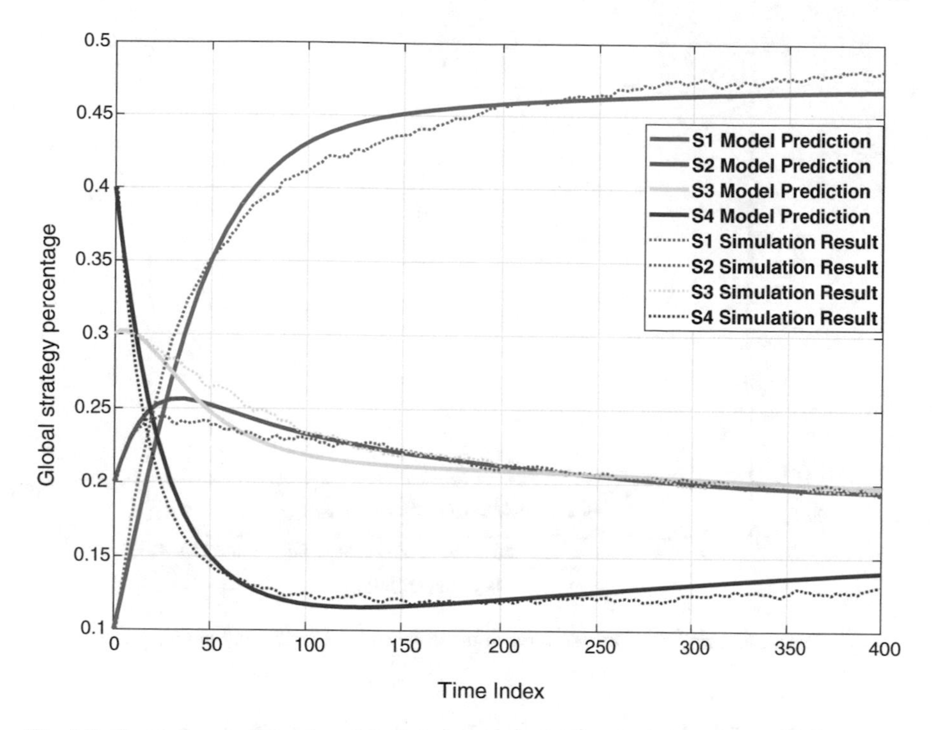

Fig. 5.8 Simulation results and model's theoretical prediction in real-world Facebook network

in Fig. 5.8. It can be seen that the theoretical results match well with the simulation results, both the diffusion curve and the evolutionary stable states, which suggestes the effectiveness of the model in real-world network structures.

5.4.2 Sina Weibo Real-World Data Experiment

In the above experiments, the parameters are first set up, then the experiments of information diffusion over different networks are conducted. In this part, we further verify the effectiveness of the model by considering a reverse process. Specifically, we use the crawled Sina Weibo (a popular Twitter-like Chinese social network) users' forwarding data to learn the payoff matrixes and other parameters. After that, we use these learned parameters and our derived multi-source information diffusion model to predict the real-world diffusion process.

In each iteration, we substitute the temporary parameters into the graphical evolutionary game model to make a diffusion estimation, which is a 4*400 matrix (four strategies and 400 units diffusion time). Therefore, the $F - norm$ of the absolute value matrix of the gap between the estimated diffusion matrix and real-data diffusion matrix can be used as a reverse parametric adjustment quantized value for the

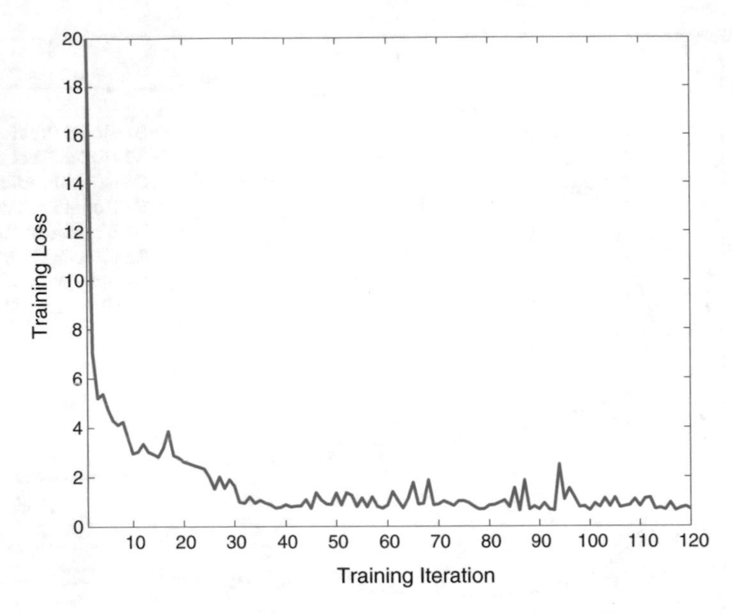

Fig. 5.9 The parameters' training loss curve in the crawled Sina Weibo real-world dataset

error function of the learning algorithm. The parameters' training loss curve can be illustrated as follows (Fig. 5.9):

We choose two pairs of popular Weibo hashtags for experiments. For each pair, there are two pieces of correlated information (all the messages in one hashtag are considered as the same information) that spread in a similar period.

The first pair is "Canada arrests Huawei's global CFO Wanzhou Meng" and "Canada former diplomatist Michael Kovrig is in detention according to the Chinese government". They have arisen large attentions in Weibo social network and became Weibo's "top search" breaking news at the same time since two hashtags are strongly relevant. Chinese people called it "It is impolite not to reciprocate" and most of them tend to forward the information in both two hashtags together. To some extent, two pieces of correlated information motivate the diffusion process each other.

The second pair is "Notre-Dame de Paris caught fire" and "Famous French game company Ubisoft announced a €500000 donation and a free download for the popular game Assassin's Creed Unity". The donation is for rebuilding and the free download game has almost the same scale of Cathedral's real-world counterpart in some scenarios. Since this game has the most sweeping and detailed buildings to ever appear in the virtual world, it may be a useful tool for rebuilding the roof of Notre-Dame de Paris cathedral. The player can also enjoy the lost heritage site. Therefore, the two pieces of information also motivate each other's diffusion processes, since different fields of people (like from the background of the art or games) may communicate together.

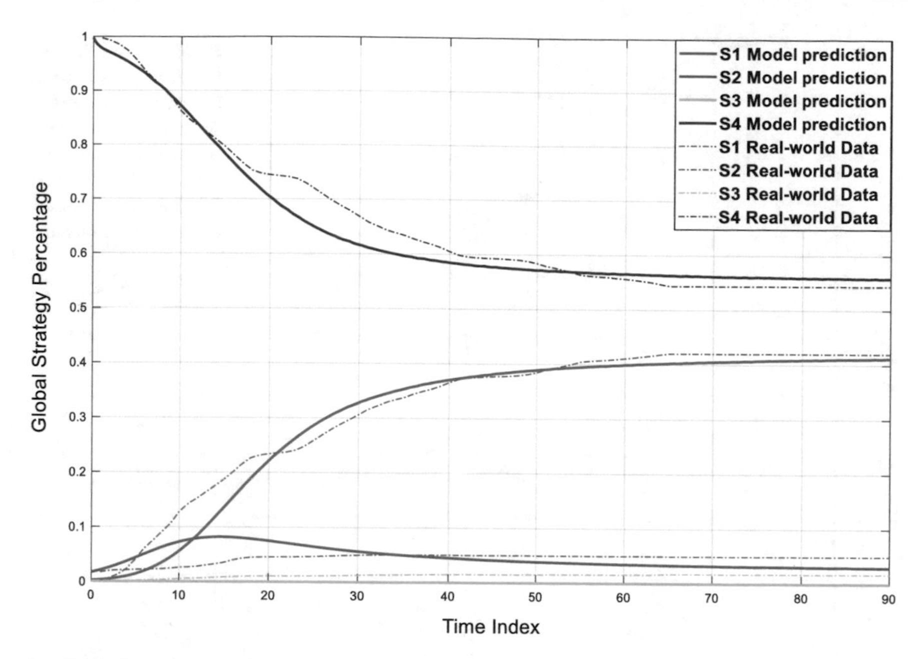

Fig. 5.10 Comparisons between the model prediction and the real-world Weibo users' forwarding data in the scenario of "Wangzhou Meng was arrested" correlated information

As we can see in Figs. 5.10 and 5.11, the model's prediction curve matches well with the real-world information diffusion process. It means that by using only a small number of parameters, our theoretical model can also fit the real information diffusion process well. The mechanism of users' decision-making and interactions are modeled, the diffused dynamics are derived and the relations of multi-source information are hidden in the payoff matrix and other parameters, with which we accomplished the desired prediction.

Finally, we conduct another experiment to verify the model's future prediction performance. An effective prediction means that we can estimate future data from the past. Accordingly, we limit the proportion of the training dataset and use the learned parameters to estimate the whole diffusion process, which is compared to the real-world diffusion data. As we can see in Fig. 5.12, when we only use 30% amounts of data for learning, the diffusion prediction does not seem reasonable. However, when only 60% ahead of training data used, we can have a good prediction performance of the whole diffusion process, which indicates our theoretical model's good performance in real-world prediction tasks.

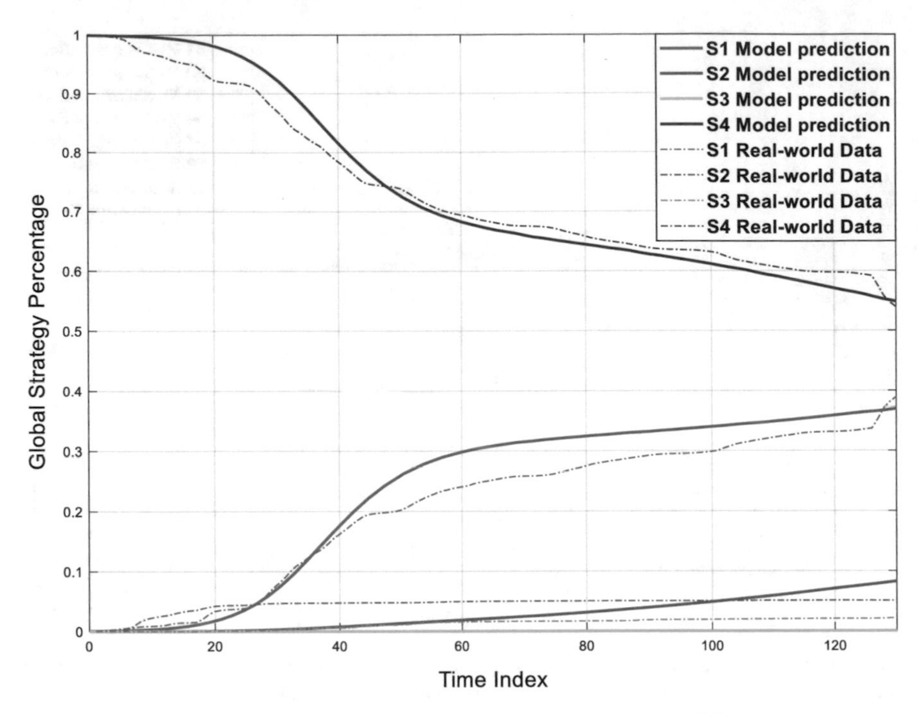

Fig. 5.11 Comparisons between the model prediction and the real-world Weibo users' forwarding data in the scenario of "Notre-Dame de Paris caught fire" correlated information

5.5 Conclusion and Future Work

Overall, this chapter proposes a new approach for multi-source information diffusion over social networks from graphical evolutionary game perspective. Through the study, we first formulate the problem by using a graphical evolutionary approach. Then, we theoretically derive three types of dynamics of the networks at different levels. Finally, we conduct different experiments to verify the correctness of our model's theoretical derivation and the effectiveness of real-world applications.

Simulation results are consistent with the proposed model and the theoretical analysis. Moreover, we further verify the effectiveness of real-world applications by using the crawled Sina Weibo users' forwarding data to learn the needed parameters. With the learned parameters, we can use our derived multi-source information diffusion model to predict different real-world diffusion processes. Moreover, our analysis and the framework are proved to have the ability to fit and predict the real-world information spreading process by learning from limited ahead amounts of data, which also demonstrates the effectiveness of the model and the analysis.

The work discussed here is just the beginning, and there are still lots of directions to explore. For example, multi-source information may interfere with each other's propagation in different ways, and it is interesting to classify and analyze different

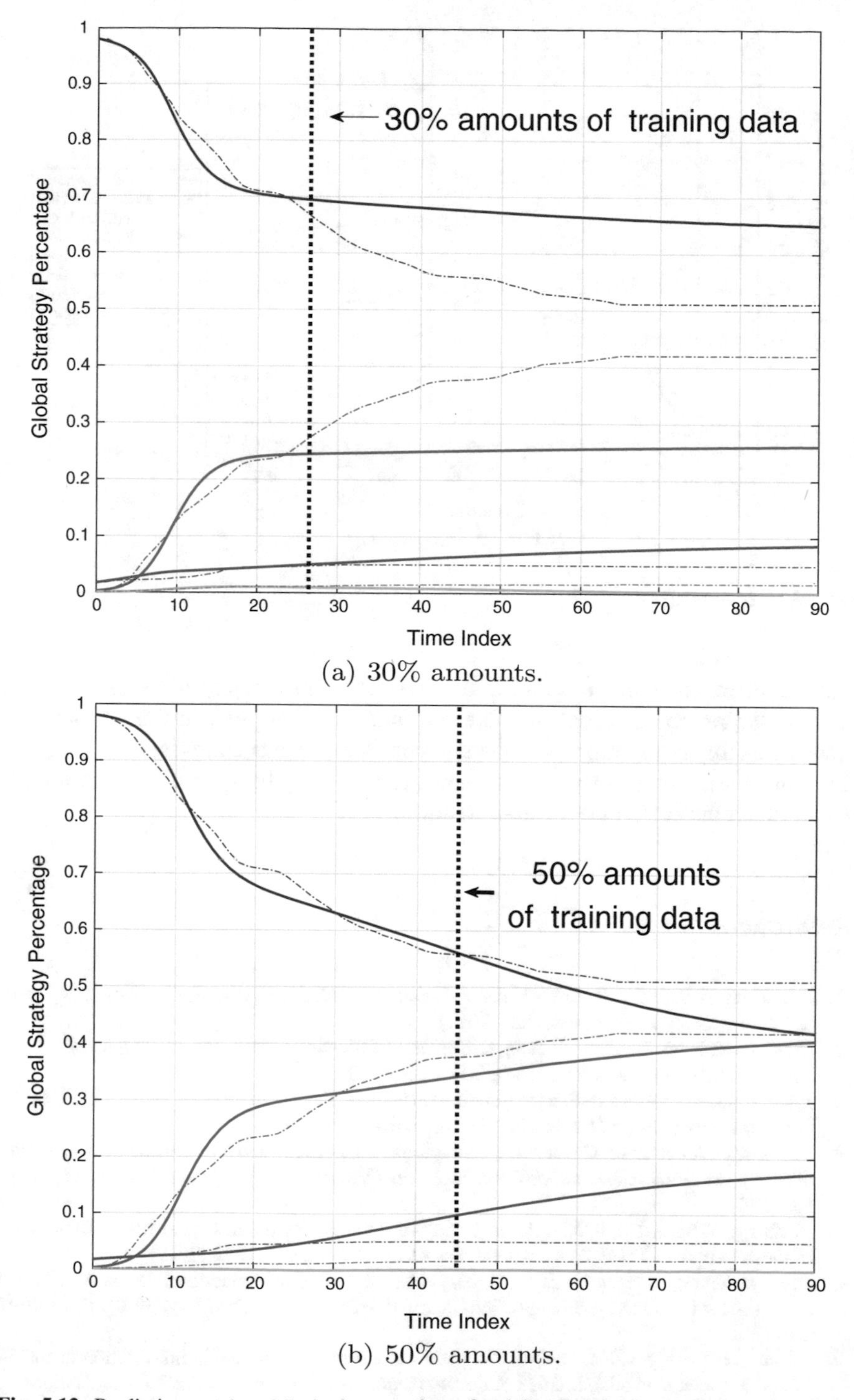

Fig. 5.12 Prediction results of limited proportion of training data in the prediction task of the scenario of "Wanzhou Meng was arrested" correlated information

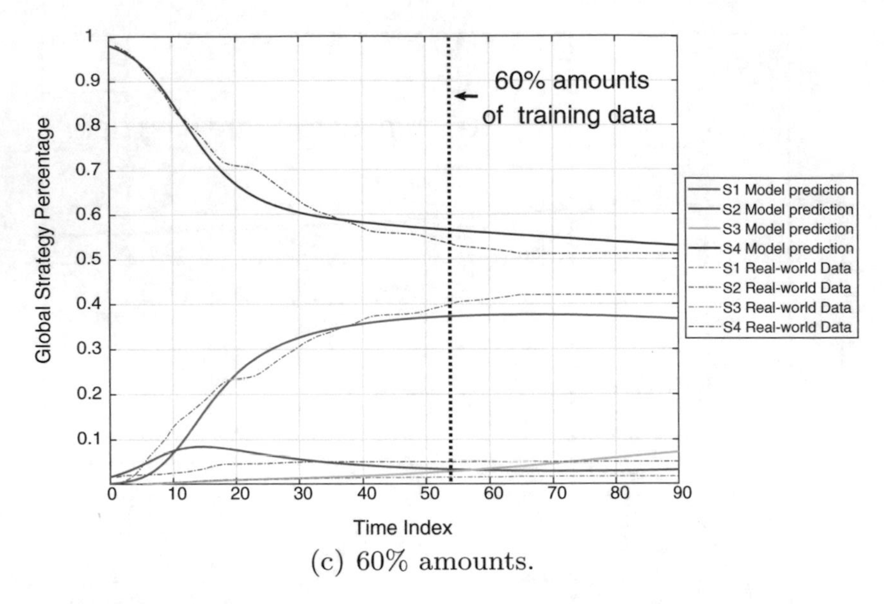

(c) 60% amounts.

Fig. 5.12 (continued)

patterns there. In addition, we need to consider more complicated scenarios, and validate the proposed model on more real data. Furthermore, in this chapter, we consider a simple scenario where the two correlated events almost start at the same time, and it is important to investigate how the time delay between the two correlated events affect the evolutionary stable states.

References

1. H.V. Zhao, W.S. Lin, K.R. Liu, *Behavior Dynamics in Media-sharing Social Networks* (Cambridge University Press, Cambridge , 2011)
2. J. Zhang, Z. Fang, W. Chen, J. Tang, Diffusion of following links in microblogging networks. IEEE Trans. Knowl. Data Eng. **27**(8), 2093–2106 (2015)
3. J. Yang, J. Leskovec, Modeling information diffusion in implicitnetworks, in *2010 IEEE International Conference on DataMining* (IEEE, 2010), pp. 599–608
4. S.A. Myers, J. Leskovec, Clash of the contagions: Cooperation and competition in information diffusion, in *IEEE International Conference on Data Mining, Conference Proceedings*, pp. 539–548
5. X. Zhang, Y. Su, S. Qu, S. Xie, B. Fang, P.S. Yu, IAD: interaction-aware diffusion framework in social networks. IEEE Trans. Knowl. Data Eng. **31**(7), 1341–1354 (2018)
6. D. Li, S. Zhang, X. Sun, H. Zhou, S. Li, X. Li, Modeling information diffusion over social networks for temporal dynamic prediction. IEEE Trans. Knowl. Data Eng. **29**(9), 1985–1997 (2017)
7. F. Wang, H. Wang, K. Xu, Diffusive logistic model towards predicting information diffusion in online social networks, in *2012 32nd International Conference on Distributed Computing Systems Workshops* (IEEE, 2012), pp. 133–139

8. J. Leskovec, M. McGlohon, C. Faloutsos, N. Glance, M. Hurst, Cascading behavior in large blog graphs, *Proceedings of SIAM International Conference on Data Mining* (2007), pp. 551–556
9. D.J. Daley, D.G. Kendall, Epidemics and rumours. Nature **204**(4963), 1118 (1964)
10. A. Beutel, B.A. Prakash, R. Rosenfeld, C. Faloutsos, Interacting viruses in networks: can both survive? in *Proceedings of the 18th ACM SIGKDD International Conference on Knowledge Discovery and Data Mining* (ACM, 2012), pp. 426–434
11. M. Granovetter, Threshold models of collective behavior. Am. J. Soc. **83**(6), 1420–1443 (1978)
12. N. Barbieri, F. Bonchi, G. Manco, Topic-aware social influence propagation models. Knowl. Inf. Syst. **37**(3), 555–584 (2013)
13. A. Guille, H. Hacid, A predictive model for the temporal dynamics of information diffusion in online social networks, in *Proceedings of the 21st International Conference on World Wide Web* (ACM, 2012), pp. 1145–1152
14. M.Y. Li, J.S. Muldowney, Global stability for the seir model in epidemiology. Math. Biosci. **125**(2), 155–164 (1995)
15. J. Yang, J. Leskovec, Modeling information diffusion in implicit networks, in *2010 IEEE International Conference on Data Mining* (IEEE, 2010), pp. 599–608
16. C. Jiang, Y. Chen, K.J.R. Liu, Graphical evolutionary game for information diffusion over social networks. IEEE J. Sel. Top. Signal Process. **8**(4), 524–536 (2014)
17. C. Jiang, C. Yan, K.J.R. Liu, Evolutionary dynamics of information diffusion over social networks. IEEE Trans. Signal Process. **62**(17), 4573–4586 (2014)
18. Y. Chen, C. Jiang, C.-Y. Wang, Y. Gao, K.R. Liu, Decision learning: data analytic learning with strategic decision making. IEEE Signal Process. Mag. **33**(1), 37–56 (2015)
19. S.J. Maynard, Evolution and the theory of games. Am. Sci. **64**(1), 41–45 (1976)
20. X. Cao, Y. Chen, C. Jiang, K.J.R. Liu, Evolutionary information diffusion over heterogeneous social networks. IEEE Trans. Signal Inf. Process. Over Netw. **2**(4), 595–610 (2016)
21. Y. Li, B. Qiu, Y. Chen, H.V. Zhao, Analysis of information diffusion with irrational users: a graphical evolutionary game approach, in *ICASSP 2019-2019 IEEE International Conference on Acoustics, Speech and Signal Processing (ICASSP)* (IEEE, 2019), pp. 2527–2531
22. Y. Chen, Y. Gao, K.R. Liu, An evolutionary game-theoretic approach for image interpolation, in *2011 IEEE International Conference on Acoustics, Speech and Signal Processing (ICASSP)* (IEEE, 2011), pp. 989–992
23. C. Jiang, Y. Chen, K.R. Liu, Distributed adaptive networks: a graphical evolutionary game-theoretic view. IEEE Trans. Signal Process. **61**(22), 5675–5688 (2013)
24. B. Wang, K.R. Liu, T.C. Clancy, Evolutionary cooperative spectrum sensing game: how to collaborate? IEEE Trans. Commun. **58**(3), 890–900 (2010)
25. J. Leskovec, Stanford large network dataset collection [online]. Available: https://snap.stanford.edu/data

Chapter 6
Analysis of Super Users in Information Diffusion

Abstract Modeling and analysis of information propagation over social networks are of great significance to better understand the avalanche of information flow and to investigate its impact on the economy and our daily life. In social networks, there exist some "super users" who have higher social status and potentially larger influence. In this chapter, we extend the graphical evolutionary game-theoretic framework to investigate their impact on information diffusion by analyzing the evolutionary dynamics and stable states. Simulation results over synthetic networks are consistent with our theoretical analysis, and demonstrate that when super users update their strategies as others do, they have little influence on the spread of information. However, when they insist on their strategies and keep forwarding the information, they have a huge impact on information propagation.

Keywords Super users · Evolutionary game theory · User heterogeneity · Information diffusion

6.1 Introduction

Nowadays, tremendous information flows are generated and propagated every single minute, due to the advances in mobile and networking technologies. Various social network platforms such as Facebook, Twitter, and Instagram have become indispensable parts of our daily life. With the popularity of such social networks, people can share their own experiences, communicate with others and fetch real-time news anywhere and anytime. On the other hand, such convenient resources have also been illegally exploited by some malicious users, which results in problems such as rumor spreading, online fraud and online bullying. Thus, it is of great importance to study different traits of social network users, investigate their impact on information diffusion, and design effective mechanisms to manage the propagation of information flows.

In the literature, tremendous efforts have been dedicated to model the information diffusion process. The existing works can be classified into two categories: data-driven approaches and model-based approaches.

Y. Chen and H. V. Zhao, *Behavior and Evolutionary Dynamics in Crowd Networks*, Lecture Notes in Social Networks, https://doi.org/10.1007/978-981-15-7160-2_6

Based on the real data generated by social network users, data-driven approaches analyzed the patterns of information dissemination over time and users' different behaviors during the diffusion process. The work in [1] argued that different information pieces exhibited various temporal patterns, that is, how information's popularity grows and fades over time. They formulated a novel time series clustering algorithm and founded that 70% information pieces exhibited spiky temporal behavior, where the peak lasted for less than 1 day. To model and predict the diffusion process without considering the network structure, the work in [2] developed the Linear Influenced Model (LIM). In LIM model, each user is estimated with an "influence function" which quantified how many other users would be influenced by him/her over time. Then, the diffusion process can be reconstructed by the influence functions. Experimented with a set of 500 million tweets and a set of 170 million news media articles, the LIM was proved to successfully predict the diffusion process.

The two above works both modeled and analyzed information diffusion in a macro view. The works in [3] studied information propagation in a micro view, and focused on social network users' forwarding behaviors. They assumed that whether a user would forward a certain piece of information is related to the previous pieces he/she has viewed. Based on this assumption, they built a model where input is an information sequence that a user browsed, and the output is the forwarding probability of the last information of the given sequence. Training and testing with real data, the authors found that the users would have a higher forwarding probability of a piece of news if he/she had browsed a related piece before. With the recent advances in deep learning techniques, the authors in [4] considered information diffusion form both micro and macro view, and proposed a novel multi-scale diffusion prediction model. The model first used recurrent neural network (RNN) to encode information diffusion state in each local area, and then predicted the future diffusion with the neural network. While improving the prediction accuracy in terms of the macroscopic diffusion, this model could also indicate which users would be infected and forwarded the information.

The data-driven methods were superior in the prediction evaluation over real-world datasets, but with less interpretability. In addition, a large number of computing resources were required due to the training of the models. Different from data-driven approaches, model-based methods relied on the understanding of mechanisms of information diffusion process and therefore had strong interpretability.

Analogous to virus spreading, a series of research works modified epidemic models to describe the information diffusion process [5–7]. Among these epidemic-based models, the "Susceptible-Infectious-Recovered" (SIR) model was the most canonical one. The SIR model classified users into different groups: those who had not heard the news (Susceptible), those who forwarded the news (Infectious) and those who stopped forwarding and lose interest in the information (Recovered). With few parameters characterizing how users become "infected" by others and "recovered", the SIR model derived dynamics equations of how ratios of different groups evolve with time. To take collective behavior and graph structure into consideration, the author in [8] assumed that a user would forward a piece of information when his/her neighbors who had forwarded it exceeded a certain threshold, and build a deterministic model, namely Linear Threshold (LT) model. To characterize different interacting strengths between users, the works in [9] proposed the Independent Cascade (IC) model. In the IC model, the relationships between users could be either "weak tie" (common relationship) or "strong tie" (closer and stronger relationship). A user would be easier to believe in the neighbors with whom he/she had strong ties, and followed their actions (forward or not forward a certain piece of information). Recently, graphical evolutionary game theory was introduced to model information diffusion [10–13]. This series of works characterized how users are influenced by their friends and forwarded the information from a game perspective. With the payoff matrix describing users' interests toward information, the diffusion process of information was derived.

A recent study showed that instead of sharing their own thoughts and experiences, social network users tended to retweet the comments of others, especially those of verified accounts and celebrities [14]. This study suggested such "super user" (or "influential user") attracted other social network users' attention, and potentially influenced their online behavior. To study the impact of such "super users", the work in [2] defined a user's influence as the number of users who were infected by him/her and shared his/her posts, and their analysis of Twitter data showed that users with more followers had a larger influence. Considering the existence of super users, the authors in [15] extended the SIR model for information diffusion. Through simulation and validation with real data, they found that super users indeed speed up the diffusion, and resulted in more users forwarding it. Similar conclusions were drawn in [16, 17], which suggested the pivotal role of super users in information dissemination. However, by simulating and examining a series of diffusion models, the authors in [18] founded that information propagation was not driven by such super users but rather a critical mass of easily influenced individuals. This observation is contrary to the most existing hypothesis of super users' influence.

In this chapter, we analyze the impact of super users based on the evolutionary game model for information diffusion [10–13]. The problem is formulated in Sect. 6.2. Then, the evolutionary dynamics of the scenario where super users use the same strategy update rule as normal users do are derived and analyzed in Sect. 6.3 and Sect. 6.4, respectively. The simulation results of this scenario are reported in Sect. 6.5.1. In Sect. 6.5.2, we discuss the simulation results of the scenario where super user keep their strategies unchanged. Finally, the conclusions are drawn in Sect. 6.6.

6.2 Problem Formulation

There are five basic elements in graphical evolutionary game theory: users, graph structure, strategy, fitness (payoff), and evolutionary stable states (ESS). To formulate information diffusion with the graphical evolutionary game framework, we need to map these five basic elements into information diffusion scenario.

Users and Graph Structure. A social network is composed of network users and the relationship among them. The structure of a social network can be abstracted as a graph where nodes represent users, and an edge between two nodes represents a certain relationship between the two corresponding users. To simplify the analysis, the network structures are assumed to be undirected and time-invariant.

- *Undirected Network.* The edges in the network are all undirected. Only when two users are friends or they have mutual and bidirectional interactions, there will be an undirected edge between the two corresponding nodes.
- *Time-invariant Network.* Both the amounts of nodes and the connections between them are assumed not to change over time.

To characterize different influences of users, two types of users are considered, namely type-1 super users and type-2 normal users. The two terms "type-1 user" and "super user", as well as "type-2 user" and "normal user" will be used interchangeably in the following of this chapter. Super users, who are often considered as the opinion leaders in the social networks, have higher social status and larger influence on others. For example, super users might refer to the verified users and celebrities accounts on Twitter. We let s be the percentage of these super users. We also assume that the super users are randomly distributed in the network.

We adopt degree distribution as the metric of the network structure. The "degree" is a term from network science, which indicates the number of edges that connect to a certain node. The degree distribution $\lambda(k)$ describes the probability of a randomly selected node (user) having k neighbors.

Strategies In the information diffusion scenario, there are two possible strategies for each user to adopt: forward the information (S_f) or do not forward it (S_n). We define p_{f1} as the percentage of super users who adopt strategy S_f, and p_{f2} as that of super users with strategy S_n, respectively. Similarly, p_{f2} and $(1 - p_{f2})$ can also be defined as the percentages of normal users with strategy S_f and S_n, respectively.

Fitness The utility function of each user is defined as fitness, which shows how well the user adapts to the surrounding environment. In the information diffusion scenario, fitness can be interpreted as the popularity of a user. Considering the difference between two types of users, the fitness of a type-i user is defined as

$$\pi_i = (1 - \alpha) \cdot b_i + \alpha \cdot U, \quad i = 1, 2. \tag{6.1}$$

Following the works in [10–13], we consider the scenario where $\alpha \ll 1$. b_i is defined as baseline fitness, and is related to users' type. Since super users have higher popularity and stronger impact on others' decision making, their baseline fitness b_1 is assumed to be greater the normal users' baseline fitness b_2. U is the total payoff received from interactions with neighbors. When two users meet, if they both forward a certain piece of information, both receive payoff u_{ff}; when they both do not forward the information, their payoffs are u_{nn}; and when they hold different strategies, both of them receive same payoff u_{fn}. The corresponding payoff matrix is

$$\begin{array}{cc} & S_f \quad S_n \\ \begin{array}{c} S_f \\ S_n \end{array} & \begin{pmatrix} u_{ff} & u_{fn} \\ u_{nf} & u_{nn} \end{pmatrix}, \end{array} \tag{6.2}$$

where a symmetric payoff matrix is considered. Here, we assume that the payoff matrix is shared by all users. Note that the payoff values depend on different kinds of information. For example, users often tend to forward the information which was recently posted and went viral online, because the forwarding of such information can attract others' attention. In such a case, the payoff values should exhibit $u_{ff} > u_{fn} > u_{nn}$. On the contrary, information like advertisements is often ignored by most of users, and the payoff value of such information should satisfy $u_{nn} > u_{fn} > u_{ff}$. Other relationships between three payoff values can be similarly interpreted.

Strategy Update Rules In social networks, whether a user would forward a certain piece of information is influenced by his/her neighbors' actions. In this sense, users' strategies may change from time to time. Intuitively, a user is more likely to forward the information when there are more neighbors who have forwarded it. We adopt the death-birth (DB) update rule from graphical evolutionary game theory to characterize such *imitation* mechanism in this section. The strategy evolution process of the whole population is divided into time units. Within each time unit, a user is randomly selected as the focal user to copy one of his/her neighbors' strategy, while others keep theirs unchanged. The probability that the focal user imitates a neighbor's strategy is proportional to that neighbor's fitness π.

Since we consider the existence of super user who are often strong-minded and tend to be less influenced by others, we also adopt the Insist (IS) rule in [13] to model this characteristics. In this chapter, we consider two different scenarios:

- *Scenario 1* Both super and normal users use the DB update rule;
- *Scenario 2* Super users use the IS rule, while normal users use the DB update rule.

Evolutionary Stable State (ESS) With the above DB strategy update rule, the whole population evolves and finally reaches the evolutionary stable state (ESS) (p_{f1}^*, p_{f2}^*). At the ESS, even if a small group of users randomly change their strategies so that the network state deviates from the ESS, it is still possible for the system to restore. In information diffusion scenario, the ESS can be used to characterize the final propagation state of the information.

6.3 Evolutionary Dynamics Analysis

In this section, we analyze the evolutionary dynamics in scenario 1, where both super users and normal users adopt the DB update rule.

With the DB update rule, the dynamics of information diffusion can be derived by analyzing the probability for a selected focal user to change his/her strategy. According to the DB update rule, the probability to select a super user as focal user is s. With probability p_{f1} and $(1 - p_{f1})$, the selected focal super user adopts strategy S_f and S_n respectively. As assumed previously that all users are randomly distributed over the network, the focal user has k neighbors with probability $\lambda(k)$. Given that the focal user has k neighbors, on average, $r = \lfloor k \cdot s \rceil$ (where $\lfloor \cdot \rceil$ is the rounding operator) of them are super users and the rest $(k - r)$ are normal users. Since the probability for a super user to adopt strategy S_f is p_{f1}, and that for a normal user is p_{f2}, the probability that k_{f1} of super user neighbors and k_{f2} of normal user neighbors adopt strategy S_f is

$$P[k_{f1}, k_{f2}|k, r] = \binom{r}{k_{f1}} \cdot p_{f1}^{k_{f1}} \cdot (1 - p_{f1})^{r-k_{f1}} \times \binom{k-r}{k_{f2}} \cdot p_{f2}^{k_{f2}} \cdot (1 - p_{f2})^{k-r-k_{f2}}. \quad (6.3)$$

According to the implementation of the DB update rule, the probability for the focal user to change his/her strategy is proportional to his/her neighbors fitness. In information diffusion scenario, a user can easily observe his/her neighbors' types (*i.e.,* whether a neighbor is a super user or a normal user) and their strategies (*i.e.,* whether a neighbor forward a certain piece of information or not). However, it is difficult for the user to know the exact fitness value of each neighbor, because the focal user has no information about his/her neighbors' neighbors, neither their types nor their adopted strategies.

Let v be the focal user, and $\{v_i\}$ be his/her k neighbors. The first step to estimate each neighbor's fitness is to find out how many neighbors v_i has. Here, we adopt *average nearest neighbors' degree (ANND)* defined in [19–21]. The average nearest neighbors' degree $m(k)$ characterizes how may neighbors v_i has, if the focal user v has k neighbors. The calculation of $m(k)$ depends on the network structure. To take the regular network as an example, all users have the same number of neighbors, k. The conclusion that $m(k) = k$ can be drawn easily. For other specific network structure, we refer the calculation of $m(k)$ to works in [19–21].

Since the focal user v has no information of v_i's neighbors, the second step to estimate v_i's fitness is to find out how many of v_i's neighbors are super users with strategy S_f and S_n, and how many are normal users using strategy S_f and S_n. According to the conclusion in [11] that the local dynamics converges faster than the global dynamics, we assume that v_i has the same distribution of neighbors with different types and strategies as v do. Specifically, v assumes that there are $M_f^1 = m(k) \cdot \frac{k_{f1}}{k}$ super users using strategy S_f, $M_n^1 = m(k) \cdot \frac{r-k_{f1}}{k}$ super users using strategy S_n, $M_f^2 = m(k) \cdot \frac{k_{f2}}{k}$ normal user using strategy S_f and the rest $M_n^2 = m(k) \cdot \frac{k-r-k_{f2}}{k}$ users are normal users with Strategy S_n. The whole process for the focal user v to estimate v_i's neighbors distribution can be summarized in Fig. 6.1.

We let $\overline{\pi}_{f1}$ be the estimated average fitness of v_i if he/she is a super user with strategy S_f. Similarly, we can also have $\overline{\pi}_{n1}$, $\overline{\pi}_{f2}$ and $\overline{\pi}_{n2}$ be the estimated average fitness when v_i is a super user with strategy S_n, a normal user with strategy S_f and a normal user with strategy S_n, respectively. With the fitness defined in (6.1) and payoff matrix in (6.2), we have

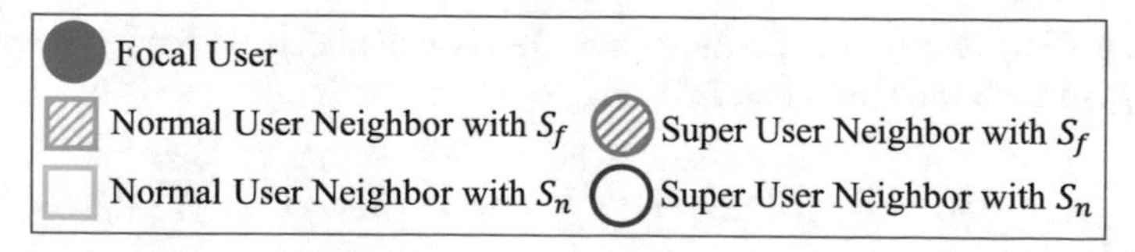

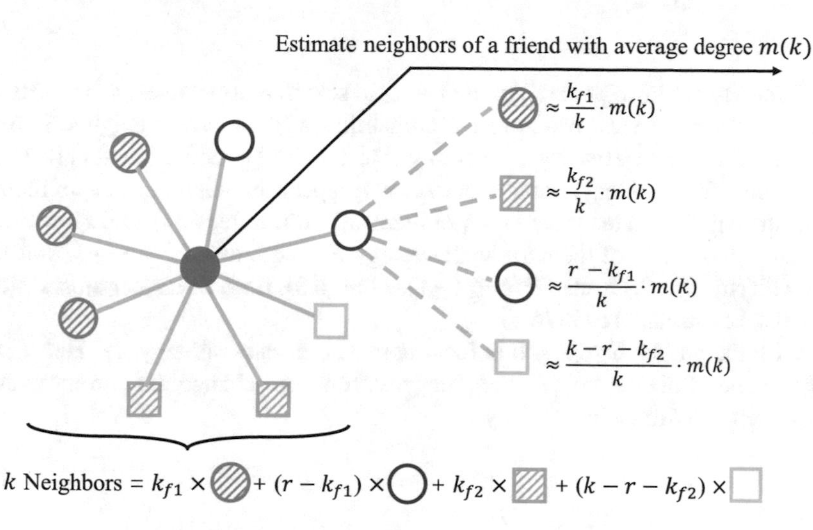

Fig. 6.1 The illustration of the focal user with neighborhood configuration (k, r, k_{f1}, k_{f2}), and how he/she estimates his/her neighbors' neighbors

$$\begin{aligned}
\overline{\pi}_{f1} &= (1-\alpha) \cdot b_1 + \alpha \cdot [(M_f^1 + M_f^2) \cdot u_{ff} + (M_n^1 + M_n^2) \cdot u_{fn}], \\
\overline{\pi}_{n1} &= (1-\alpha) \cdot b_1 + \alpha \cdot [(M_f^1 + M_f^2) \cdot u_{fn} + (M_n^1 + M_n^2) \cdot u_{nn}], \\
\overline{\pi}_{f2} &= (1-\alpha) \cdot b_2 + \alpha \cdot [(M_f^1 + M_f^2) \cdot u_{ff} + (M_n^1 + M_n^2) \cdot u_{fn}], \\
\overline{\pi}_{n2} &= (1-\alpha) \cdot b_2 + \alpha \cdot [(M_f^1 + M_f^2) \cdot u_{fn} + (M_n^1 + M_n^2) \cdot u_{nn}].
\end{aligned} \tag{6.4}$$

If the focal super user is with strategy S_f and decides to copy the strategy of a neighbor who users S_n, according to DB update rule, this happens with probability

$$P_{f1 \to n1} = \frac{(r - k_{f1}) \cdot \overline{\pi}_{n1} + (k - r - k_{f2}) \cdot \overline{\pi}_{n2}}{k_{f1} \cdot \overline{\pi}_{f1} + k_{f2} \cdot \overline{\pi}_{f2} + (r - k_{f1}) \cdot \overline{\pi}_{n1} + (k - r - k_{f2}) \cdot \overline{\pi}_{n2}}. \tag{6.5}$$

This strategy update results in p_{f1} decreasing by $1/(N \cdot s)$. Similarly, the probability that focal super user changes his/her strategy from S_n to S_f is

$$P_{n1 \to f1} = \frac{k_{f1} \cdot \overline{\pi}_{f1} + k_{f2} \cdot \overline{\pi}_{f2}}{k_{f1} \cdot \overline{\pi}_{f1} + k_{f2} \cdot \overline{\pi}_{f2} + (r - k_{f1}) \cdot \overline{\pi}_{n1} + (k - r - k_{f2}) \cdot \overline{\pi}_{n2}}, \tag{6.6}$$

and it increases p_{f1} by $1/(N \cdot s)$.

Summarizing (6.5) and (6.6), the dynamics $\dot{p}_{f1}$ is calculated as the expected variation of p_{f1} in each time unit, that is,

$$\dot{p}_{f1} = \frac{s}{N \cdot s} \cdot \sum_{k,k_{f1},k_{f2}} \lambda(k) \cdot P[k_{f1}, k_{f2}|k, r] \times [(1 - p_{f1}) \cdot P_{n1 \to f1} - p_{f1} \cdot P_{f1 \to n1}]. \tag{6.7}$$

Similarly, we have probability of $(1 - s)$ to select a normal user as the focal user. This focal normal user also have the probability $\lambda(k)$ to have k neighbors. As we have assumed that all users are randomly distributed all over the social network, the probability for this focal normal user to have r super user neighbors (k_{f1} of them are with strategy S_f and the rest $r - k_{f1}$ of them are with strategy S_n), and $k - r$ normal user neighbors (k_{f2} of them are with strategy S_f and the rest $k - r - k_{f2}$ of them are with strategy S_n) is also $P(k_{f1}, k_{f2}|k, r)$ in (6.3). Each of his neighbors' fitness can also be estimated as in (6.4).

With the probability p_{f2}, this focal normal user uses strategy S_f. He/she may imitate one of his/her neighbors strategy with S_n and change the current strategy. This happens with the probability

$$P_{f2 \to n2} = P_{f1 \to n1}. \tag{6.8}$$

This strategic changing cause p_{f2} to decrease by $1/(N \cdot (1 - s))$. Similarly, with probability $(1 - p_{f2})$, the focal normal user uses strategy S_n. The probability for him/her to copy one of his/her neighbors with strategy S_f and change the current strategy to S_f is

$$P_{n2 \to f2} = P_{n1 \to f1}. \tag{6.9}$$

This results in p_{f2} increasing by $1/(N \cdot (1 - s))$. Thus, the dynamics of p_{f2} is

$$\begin{aligned} \dot{p}_{f2} =& \frac{(1 - s)}{N \cdot (1 - s)} \\ & \times \sum_{k,k_{f1},k_{f2}} \lambda(k) \cdot P[k_{f1}, k_{f2}|k, r] \times [(1 - p_{f2}) \cdot P_{n2 \to f2} - p_{f2} \cdot P_{f2 \to n2}]. \end{aligned} \tag{6.10}$$

6.4 ESS Analysis

Given the derived evolutionary dynamics equations in (6.7) and (6.10), in order to obtain the ESS (p^*_{f1}, p^*_{f2}), we first derive the equilibrium states by letting the two dynamics equations equals to 0, that is,

$$\dot{p}_{f1} = 0, \text{ and } \dot{p}_{f2} = 0, \tag{6.11}$$

Then, according to definition of stable state in [10], the ESS are the derived equilibrium states which satisfy

$$\frac{\partial \dot{p}_{f1}}{\partial p_{f1}} + \frac{\partial \dot{p}_{f2}}{\partial p_{f2}} < 0, \text{ and } \frac{\partial \dot{p}_{f1}}{\partial p_{f1}} \cdot \frac{\partial \dot{p}_{f2}}{\partial p_{f2}} - \frac{\partial \dot{p}_{f1}}{\partial p_{f2}} \cdot \frac{\partial \dot{p}_{f2}}{\partial p_{f1}} > 0. \tag{6.12}$$

The solution to (6.11) and (6.12) is

$$p_{f1}^* = p_{f2}^* = \begin{cases} 0, & \text{if } u_{nn} > \max(u_{fn}, u_{ff}), \\ 1, & \text{if } u_{ff} > \max(u_{fn}, u_{nn}), \\ -\dfrac{\Phi \cdot (\overline{m/k} - \overline{m/k^2}) + \Phi_n \cdot (\overline{m} - \overline{m/k})}{\Phi \cdot [\overline{m} - 3 \cdot \overline{m/k} + 2 \cdot \overline{m/k^2}]}, & \text{if } u_{fn} > \max(u_{ff}, u_{nn}), \end{cases} \tag{6.13}$$

where $\Phi_n = u_{fn} - u_{nn}, \Phi = u_{ff} - 2 \times u_{fn} + u_{nn}, \overline{m/k^2} = \sum_k \lambda(k) \cdot \frac{m(k)}{k^2}, \overline{m/k} = \sum_k \lambda(k) \cdot \frac{m(k)}{k}$ and $\overline{m} = \sum_k \lambda(k) \cdot m(k)$. The detailed derivation can be found in the Appendix.

From (6.13), we have the following observations:

- $p_{f1}^* = p_{f2}^*$. That is, the percentage of users using strategy S_f among super users and that among normal users are the same at the ESS.
- p_{f1}^* and p_{f2}^* do not depend on the baseline fitness of super user b_1, that of normal user b_2, and the percentage of super users s. That is, the evolutionary stable state is not influenced by super users. An intuitive explanation of this observation is that super users use the same strategy update rule as normal users, and they are influenced by normal users in the same way that normal users are influenced by them.
- Finally, when u_{ff} is the largest, the information is related with recent hot topics. Forwarding the information can bring a user more popularity and attention from others. From (6.13), both p_{f1}^* and p_{f2}^* equal to 1 in this scenario, and all users would forward the information since forwarding can obtain higher payoffs. On the contrary, when the information is of less interests, u_{nn} is the largest. In this case, not forwarding brings higher payoffs and no one will forward it. In all other cases, the ESS is between 0 and 1.

6.5 Simulation Results

In this section, we run simulations on synthetic networks in the two different scenarios in 6.2 to validate the analysis and argue the influence of super users.

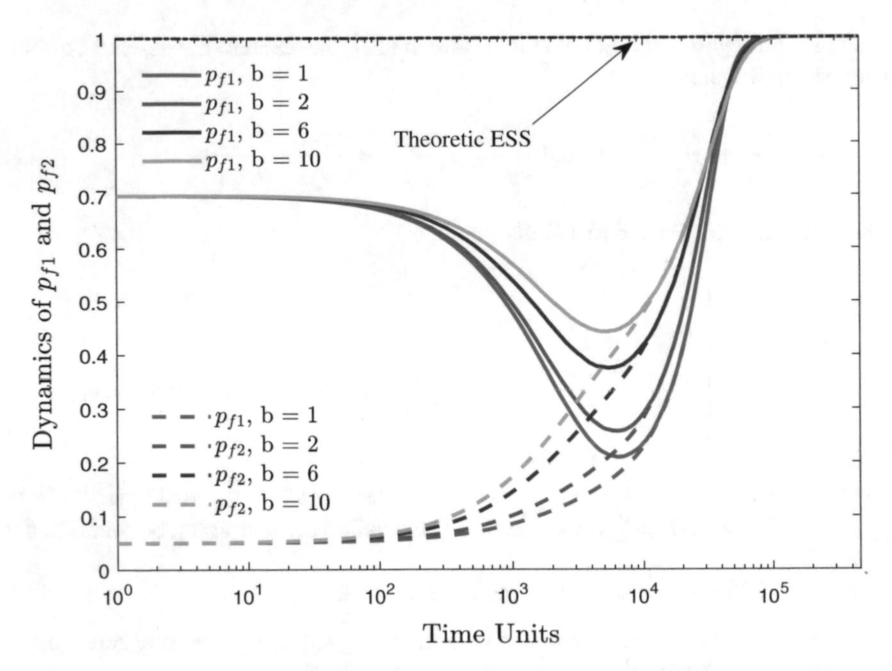

Fig. 6.2 Evolutionary dynamics of p_{f1} and p_{f2} in scenario 1 with payoff matrix PM_1 and $s = 0.1$ over BA scale-free networks (k = 40)

6.5.1 Simulation of Scenario 1

To validate the analysis result of scenario 1, we run simulations on Barabási-Albert (BA) scale-free networks. Here, we assume that there are $N = 2000$ users in the network, and the average degree of the network is $\bar{k} = 40$. We let $\alpha = 0.01$ and $b_2 = 1$. We choose the following three different payoff values:

- $\mathbf{PM_1}$: $u_{ff} = 0.9, u_{fn} = 0.4, u_{nn} = 0.2$;
- $\mathbf{PM_2}$: $u_{ff} = 0.4, u_{fn} = 0.9, u_{nn} = 0.2$; and
- $\mathbf{PM_3}$: $u_{ff} = 0.2, u_{fn} = 0.4, u_{nn} = 0.9$,

which corresponds to three different scenarios in the analysis of ESS results in (6.13). The initial values of p_{f1} and p_{f2} are set to 0.7 and 0.05, respectively. For each simulation setting, we randomly generated 10 networks, and conduct 80 simulation runs for each network. In each simulation run, the strategy update process is repeated until the network reaches the stable state.

We first fix $s = 0.1$, that is, there are 200 super users randomly distributed all over the synthetic network, and change the baseline fitness of super users $b_1 = \{1, 2, 6, 10\}$ and conduct the DB simulation process on the synthetic network. The results of three different payoff matrix are in Figs. 6.2, 6.3, and 6.4. From these results, we can have the following observations:

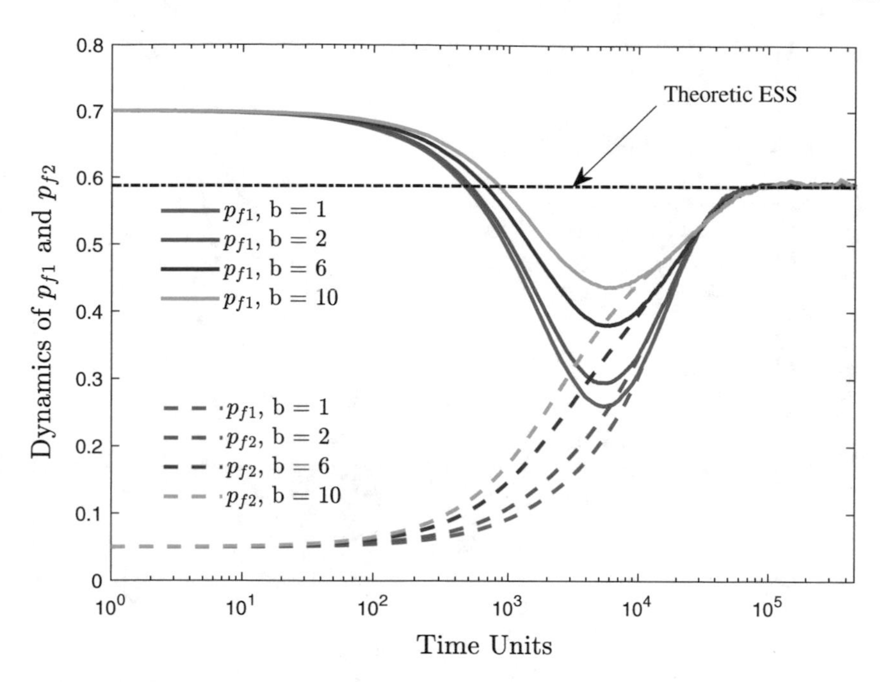

Fig. 6.3 Evolutionary dynamics of p_{f1} and p_{f2} in scenario 1 with payoff matrix PM_2 and $s = 0.1$ over BA scale-free networks (k = 40)

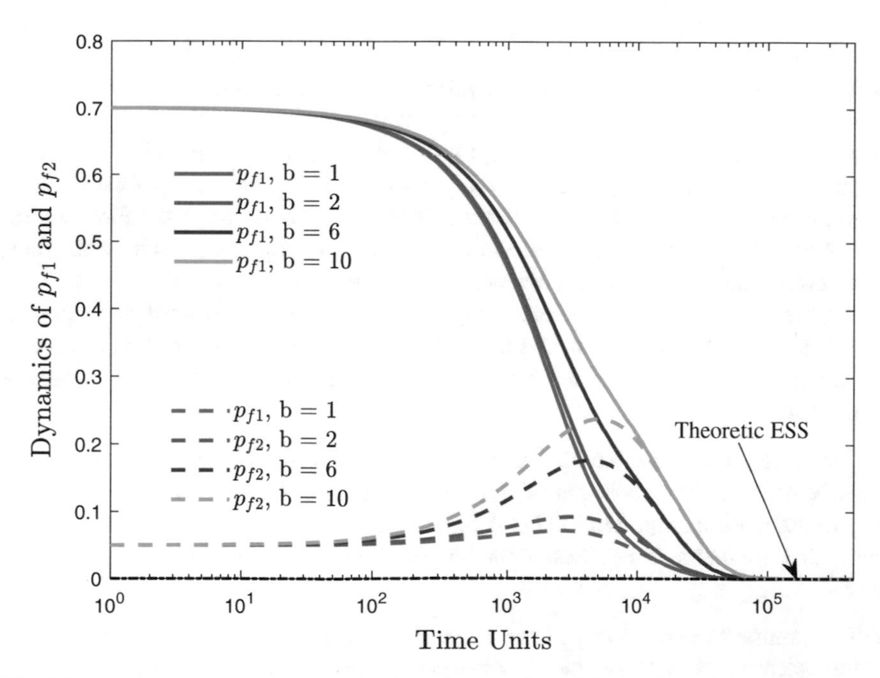

Fig. 6.4 Evolutionary dynamics of p_{f1} and p_{f2} in scenario 1 with payoff matrix PM_3 and $s = 0.1$ over BA scale-free networks (k = 40)

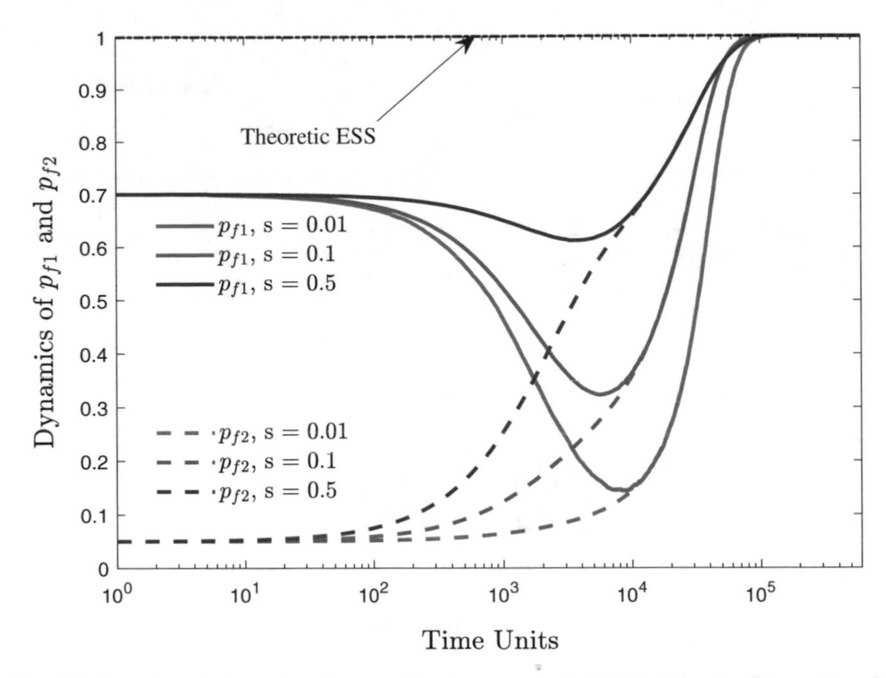

Fig. 6.5 Evolutionary dynamics of p_{f1} and p_{f2} in scenario 1 with payoff matrix PM_1 and $b_1 = 4$ over BA scale-free networks (k = 40)

- The predicted ESS given by the analysis in (6.13) match well with the simulation results.
- The ESS are the same for all different values of super users' fitness b_1 when other parameters are the same, even when $b_1 = b_2 = 1$ (there are no super users exists in the networks). This suggests that even super users have larger baseline fitness, and potentially they can influence more neighbors to copy their strategy, the final spread of the information is not determined by this factor.
- When b_1 increases, the evolution of p_{f1} becomes slower, while that of p_{f2} becomes faster. This observation indicates that the baseline fitness of super users can speed up the strategic evolution among normal users, that is, the evolution process is influenced by b_1.

Next, we fix the baseline fitness of super users $b_1 = 4$ and varies the percentage of super users s. The evolution of p_{f1} and p_{f2} with different value of s and three payoff matrices are reported in Figs. 6.5, 6.6, and 6.7. From the results, we can also show that the theoretic ESS match the simulation results. Besides, we can also see that

- The simulation process of p_{f1} and p_{f2} with different percentages of super users s finally converge to the same ESS, when other parameters are the same. This shows that no matter how many super users exists in the network, the final spread of the

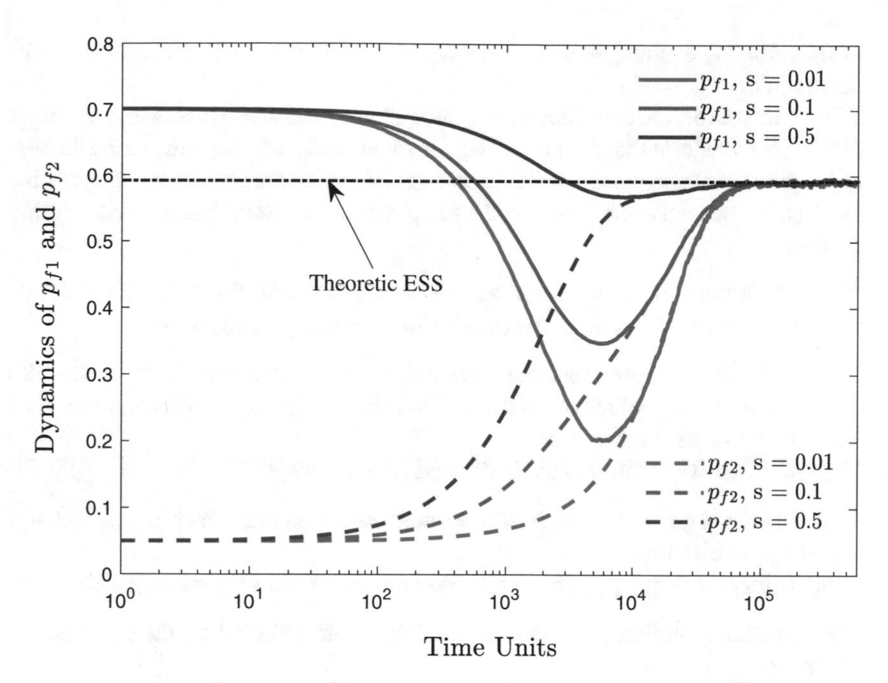

Fig. 6.6 Evolutionary dynamics of p_{f1} and p_{f2} in scenario 1 with payoff matrix PM_2 and $b_1 = 4$ over BA scale-free networks (k = 40)

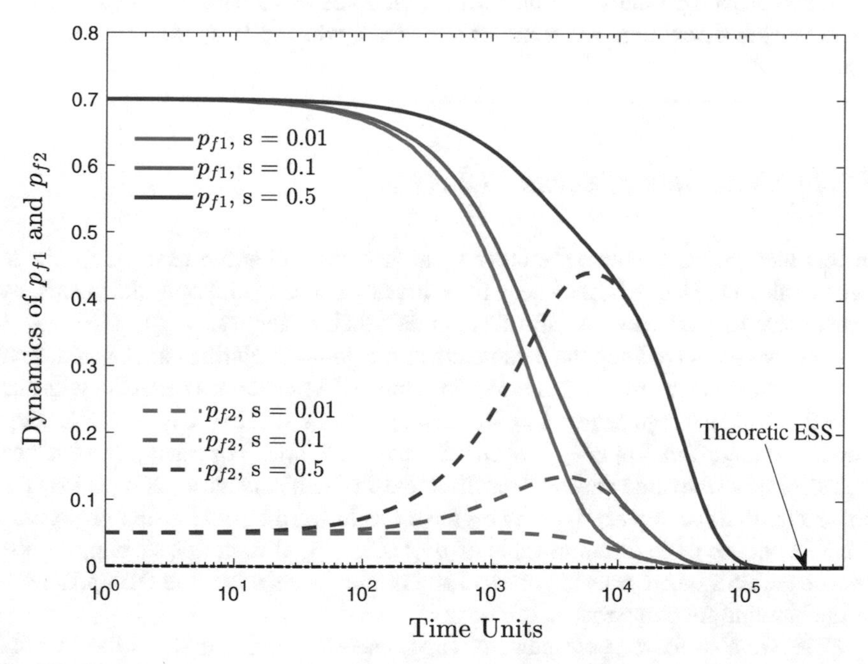

Fig. 6.7 Evolutionary dynamics of p_{f1} and p_{f2} in scenario 1 with payoff matrix PM_3 and $b_1 = 4$ over BA scale-free networks (k = 40)

information is not changed when both super users and normal users adopt the DB update rule.
- When s increases, the evolution of p_{f1} becomes slower, while that of p_{f2} becomes faster. This suggests that the percentage of super users influence the strategic evolution process of normal users. The more super users there exists in the network, the more easily normal users influenced by super users' strategy during the evolution process.

Combining the simulation results and the observations in Figs. 6.2, 6.3, 6.4, 6.5, 6.6, and 6.7, we can summarize the conclusion of scenario 1 as follow:

- The simulation ESS are the same with and without super users, that is, the ESS are the same for all values of b_1 and s. This observation agree with our theoretic analysis in the previous section.
- The the entire evolution process is divided into two stages:
  - In the first stage, super users and normal users influence each other, and p_{f1} and p_{f2} evolve until $p_{f1} = p_{f2}$;
  - In the second stage, p_{f1} and p_{f2} evolve together until they reach the ESS.

  The strategic evolution process of normal users are infected by the existence of super users.

In information diffusion scenario, we could not deny that super users' tweets are often forwarded by others. In other words, their strategies barely change. Thus, in the following discussion, we consider scenario 2 where all super users use the IS rule.

6.5.2 Simulation of Scenario 2

In this section, we simulate the other scenario where all super users adopt the IS update rule in [13], that is, they keep their strategy unchanged. For normal users, we assume that they still use the DB rule to update their strategies.

Here, we intend to keep the parameters in previous simulations unchanged. That is, the network structure remains Barabási-Albert (BA) scale-free networks with size $N = 2000$ and average degree $\overline{k} = 40$. α and b_2 are also set to 0.01 and 1. We adopt the second payoff matrix PM_2 in Sect. 6.5.1 as an example. For other payoff values, we can observe the same trend and results. Note that in this scenario, as super users do not change their strategies, p_{f1} remains the same during the entire evolution process. We here choose three different value of p_{f1}: $0.3,\ 0.5,\ 0.7$. In this section, we also choose the ESS where both super users and normal users adopt the DB update rule as the baseline for comparison.

First, We also fix the percentage of super users $s = 0.1$ and varies the baseline fitness of super users b_1. The simulation ESS results is shown in Fig. 6.8. Compared with the baseline simulation results, the ESS no longer equals to the scenario where

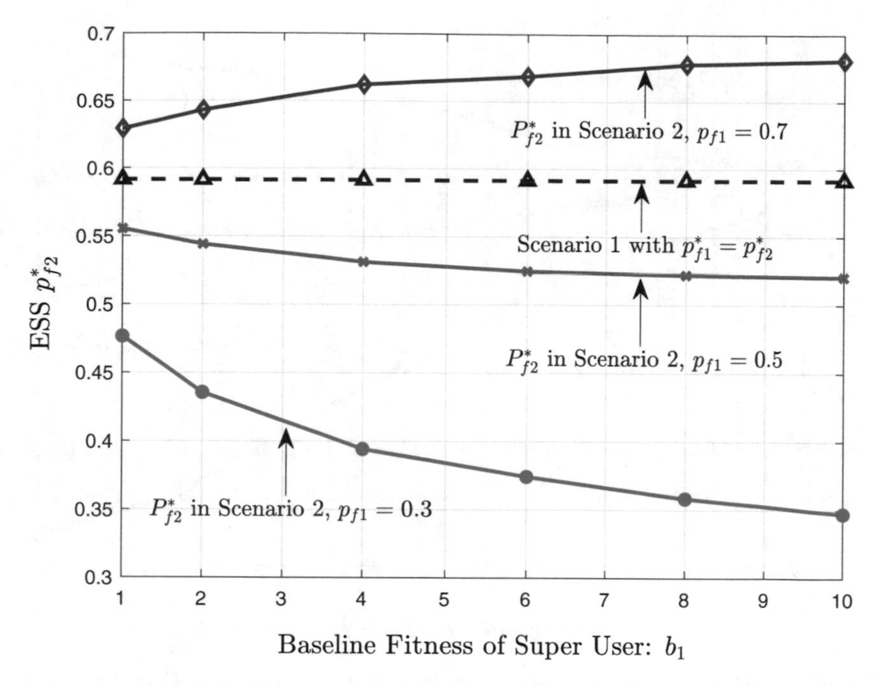

Fig. 6.8 The ESS (p^*_{f1}, p^*_{f2}) in Scenario 2 with payoff matrix PM_2 and $b_1 = 4$ over BA scale-free networks (k = 40)

no super users exist, and it also changes as the variation of baseline fitness of super users b_1. This suggests that the super users have great influence on other normal users when they keep their strategies unchanged. We can also show that p^*_{f2} get closer to the previous set p_{f1} as b_1 increased. The intuition of this observation is that when super users have a larger baseline fitness (they have larger influence on others decision making), normal users are more easily influenced by them and more likely to copy the super user neighbors' strategies. As assumed that these super users are randomly distributed all over the network, the percentage of normal users that are influenced by super users with strategy S_f and also forward the information also takes roughly the percentage of p_{f1} among all normal users.

Next, we fix the baseline fitness of super users $b_1 = 4$ and vary the percentage of super users s. The simulation ESS results are shown in Fig. 6.9. We can show when there are more users exists in the network, the ESS of p^*_{f2} deviates more compared to the baseline scenario. That is, when there are more super users, normal users are more likely to meet with super users, influence by them and copy their strategies.

In summary, super users greatly influence the spread of information among normal users when they keep their strategies unchanged. When there are more super users and when their baseline fitness b_1 are larger, p^*_{f2} at the ESS is closer to p_{f1}, and more normal users are influenced by super users and copy their strategies. Compared with the first scenario where all users adopt DB rule, we observe that strategy update

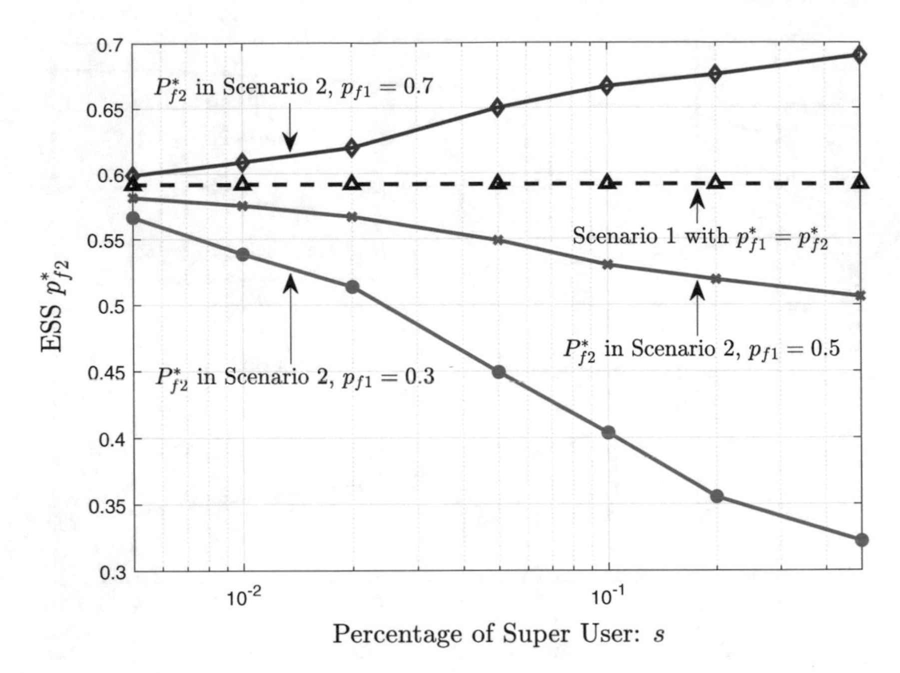

Fig. 6.9 The ESS (p^*_{f1}, p^*_{f2}) in Scenario 2 with payoff matrix PM_2 and $b_1 = 4$ over BA scale-free networks (k = 40)

rule plays an important role in information diffusion, and super users influence the stable states only if they keep their strategies fixed. This agrees with the observations in [18].

6.6 Conclusion and Future Work

In this chapter, we theoretically analyze the influence of super users on the spread of information by extending the graphical evolutionary game framework for information diffusion. We analyze the evolutionary dynamics and the stable states of information diffusion when super users use the DB and the IS update rules, respectively. Our investigation shows that super users themselves can not change the evolutionary stable state of information diffusion, but only the evolutionary dynamics; while strategy update rules are critical to the evolutionary stable states of the diffusion process. This analysis is crucial to the understanding of the evolutionary dynamics of information diffusion over social networks.

Future work includes validation of our analysis on real Weibo or Twitter data, and to investigate the impact of such "influential users" or "super users" on public opinion. Another interesting future research direction is to jointly consider different types of "heterogeneity", including users' social status (baseline fitnees), their strategy updating rules, and different types of payoff matrices, and analyze their impact on the network.

Appendix

In this section, we elaborate the derivation of the ESS in (6.13). We first denote

$$\gamma_f = \frac{m(k)}{k} \times \left[(k_{f1} + k_{f2}) \cdot u_{ff} + (k - k_{f1} - k_{f2}) \cdot u_{fn}\right], \text{ and} \tag{6.14}$$

$$\gamma_n = \frac{m(k)}{k} \times \left[(k_{f1} + k_{f2}) \cdot u_{fn} + (k - k_{f1} - k_{f2}) \cdot u_{nn}\right], \tag{6.15}$$

respectively, for annotation simplification. Taking the strategic changing probability in (6.5) and (6.6) into the dynamics of p_{f1} in (6.7), we have

$$\begin{aligned}
\dot{p}_{f1} &= \frac{1}{N} \cdot \sum_{k,k_{f1},k_{f2}} \lambda(k) \cdot P[k_{f1}, k_{f2}|k, r] \times [(1 - p_{f1}) \cdot P_{n1\to f1} - p_{f1} \cdot P_{f1\to n1}] \\
&= \frac{1}{N} \cdot \mathrm{E}\left\{(1 - p_{f1}) \cdot P_{n1\to f1} - p_{f1} \cdot P_{f1\to n1}\right\} \\
&= \frac{1}{N} \cdot \mathrm{E}\left\{\frac{x_1 + \alpha \cdot x_2}{x_3 + \alpha \cdot x_4}\right\},
\end{aligned} \tag{6.16}$$

where $\mathrm{E}\{\cdot\}$ represents taking expectation with respect to k, k_{f1} and k_{f2}. x_1, x_2, x_3 and x_4 respectively are

$$\begin{aligned}
x_1 =& b_1(k_{f1} - r \cdot p_{f1}) - b_2[(k - r) \cdot p_{f1} - k_{f2}], \\
x_2 =& [k_{f1}(\gamma_f - b_1) + k_{f2}(\gamma_f - b_2)](1 - p_{f1}) \\
& - p_{f1}[(\gamma_n - b_1)(r - k_{f1}) + (\gamma_n - b_2)(k - r - k_{f2})], \\
x_3 =& b_1 \cdot r + b_2 \cdot (k - r), \quad \text{and} \\
x_4 =& (\gamma_n - b_1) \cdot (r - k_{f1}) + k_{f2} \cdot (\gamma_f - b_2) \\
& + k_{f1} \cdot (\gamma_f - b_1) + (\gamma_n - b_2) \cdot (k - r - k_{f2}).
\end{aligned} \tag{6.17}$$

With the assumption that $\alpha \ll 1$, we can use the approximation $\frac{x_1+\alpha\cdot x_2}{x_3+\alpha\cdot x_4} \approx \frac{x_1}{x_3} + \alpha \times \frac{x_2\cdot x_3 - x_1\cdot x_4}{x_3^2}$ to simplify the term in expectation. We also use $k \cdot s$ to approximate r, because k is often assumed to be large in social networks. Thus, we can further obtain

$$
\begin{aligned}
&\mathrm{E}\left\{\frac{x_1+\alpha\cdot x_2}{x_3+\alpha\cdot x_4}\right\}\approx \mathrm{E}\left\{\frac{x_1}{x_3}+\alpha\times\frac{x_2\cdot x_3-x_1\cdot x_4}{x_3^2}\right\}\\
&=\sum_k\lambda(k)\left[\sum_{k_{f1}=0}^{r}\sum_{k_{f2}=0}^{k-r}P[k_{f1},k_{f2}|k,r]\left(\frac{x_1}{x_3}+\alpha\times\frac{x_2\cdot x_3-x_1\cdot x_4}{x_3^2}\right)\right]\\
&=\frac{b_2\cdot(p_{f2}-p_{f1})\cdot(1-s)}{b_1\cdot s+b_2\cdot(1-s)}\\
&+\alpha\cdot\sum_k\lambda(k)\cdot\left[\frac{m(k)}{k^2}\cdot\frac{f_1+f_2\times k\cdot m(k)+f_3\times k^2\cdot m(k)}{[b_1\cdot s+b_2\cdot(1-s)]^2}\right],
\end{aligned}
\tag{6.18}
$$

where f_1, f_2 and f_3 respectively are

$$
\begin{aligned}
f_1=\ &\Phi\cdot b_1\cdot p_{f1}\cdot s\cdot(1-p_{f1})\cdot(2p_{f1}-1)\\
&+\Phi\cdot b_2\cdot p_{f2}\cdot(1-s)\cdot(1-p_{f2})\cdot(2p_{f1}-1),\\
f_2=\ &b_1\cdot s\cdot[-p_{f2}^2\cdot((p_{f1}-1)\cdot\Phi-\Phi_n)\cdot(s-1)\\
&-p_{f2}\cdot(\Phi-3p_{f1}\cdot\Phi+2p_{f1}^2\cdot\Phi+\Phi_n)\cdot(s-1)\\
&+(p_{f1}-1)\cdot p_{f1}\cdot(-\Phi_n\cdot(s-2)+(3p_{f1}-1)\cdot\Phi\cdot s)]\\
&+b_2\cdot(s-1)\cdot[3p_{f2}^3\cdot\Phi\cdot(s-1)+(p_{f1}-1)\cdot p_{f1}\cdot(\Phi+\Phi_n)\cdot s\\
&+p_{f2}\cdot(\Phi_n\cdot(1+s)+\Phi\cdot(s-1+3p_{f1}\cdot s-p_{f1}^2\cdot s))\\
&-p_{f2}^2\cdot(\Phi_n\cdot(1+s)+2\Phi\cdot((2+p_{f1})\cdot s-2))],\ \text{and}\\
f_3=\ &b_1\cdot s\cdot[-p_{f2}^2\cdot((p_{f1}-1)\cdot\Phi-\Phi_n)\cdot(s-1)^2\\
&+p_{f2}\cdot(s-1)\cdot(2p_{f1}^2\cdot\Phi\cdot s-2p_{f1}\cdot(\Phi_n\cdot(s-1)+\Phi\cdot s)-u_{fn})\\
&+p_{f1}(\Phi_n+p_{f1}\cdot\Phi_n\cdot(s-2)\cdot s+(1-p_{f1})\cdot p_{f1}\cdot\Phi\cdot s^2+(s-1)\cdot u_{fn})]\\
&+b_2\cdot(s-1)\cdot[p_{f2}^3\cdot\Phi\cdot(s-1)^2\\
&-p_{f2}^2\cdot(s-1)\cdot(\Phi_n\cdot(1+s)+\Phi\cdot(s-1+2\cdot p_{f1}\cdot s))\\
&-p_{f1}\cdot s\cdot(p_{f1}\cdot(\Phi+\Phi_n)\cdot s+u_{fn})\\
&+p_{f2}\cdot(\Phi_n\cdot(2p_{f1}\cdot s^2-1)+s\cdot(2p_{f1}\cdot\Phi\cdot(s-1)+p_{f1}^2\cdot\Phi\cdot s+u_{fn}))].
\end{aligned}
\tag{6.19}
$$

Here, $\Phi=u_{ff}-2\cdot u_{fn}+u_{nn}$ and $\Phi_n=u_{fn}-u_{nn}$. We also let

$$
\overline{m/k^2}=\sum_k\lambda(k)\cdot\frac{m(k)}{k^2},\quad \overline{m/k}=\sum_k\lambda(k)\cdot\frac{m(k)}{k},\quad\text{and}\quad \overline{m}=\sum_k\lambda(k)\cdot m(k).
\tag{6.20}
$$

Consequently, we can have

$$
\dot{p}_{f1}=\frac{b_2\cdot(p_{f2}-p_{f1})\cdot(1-s)}{N\cdot[b_1\cdot s+b_2\cdot(1-s)]}+\alpha\times\frac{f_1\cdot\overline{m/k^2}+f_2\cdot\overline{m/k}+f_3\cdot\overline{m}}{N\cdot[b_1\cdot s+b_2\cdot(1-s)]^2}.
\tag{6.21}
$$

With the similar derivation, we can also have the dynamics of p_{f2}, that is

$$\dot{p}_{f2} = \frac{b_1 \cdot (p_{f1} - p_{f2}) \cdot s}{N \cdot [b_1 \cdot s + b_2 \cdot (1-s)]} + \alpha \times \frac{f_1 \cdot \overline{m/k^2} + f_2 \cdot \overline{m/k} + f_3 \cdot \overline{m}}{N \cdot [b_1 \cdot s + b_2 \cdot (1-s)]^2}. \quad (6.22)$$

To calculate the ESS, we first need to let $\dot{p}_{f1}$ and $\dot{p}_{f2}$ equal to zero to obtain equilibrium points. That is,

$$\dot{p}_{f1} = \frac{b_2 \cdot (p_{f2} - p_{f1}) \cdot (1-s)}{N \cdot [b_1 \cdot s + b_2 \cdot (1-s)]} + \alpha \times \frac{f_1 \cdot \overline{m/k^2} + f_2 \cdot \overline{m/k} + f_3 \cdot \overline{m}}{N[b_1 \cdot s + b_2 \cdot (1-s)]^2} = 0, \quad \text{and} \quad (6.23)$$

$$\dot{p}_{f2} = \frac{b_1 \cdot (p_{f1} - p_{f2}) \cdot s}{N \cdot [b_1 \cdot s + b_2 \cdot (1-s)]} + \alpha \times \frac{f_1 \cdot \overline{m/k^2} + f_2 \cdot \overline{m/k} + f_3 \cdot \overline{m}}{N[b_1 \cdot s + b_2 \cdot (1-s)]^2} = 0. \quad (6.24)$$

Combining the two equations above, we can have

$$\frac{b_2 \cdot (p_{f2} - p_{f1}) \cdot (1-s)}{N \cdot [b_1 \cdot s + b_2 \cdot (1-s)]} + \alpha \times \frac{f_1 \cdot \overline{m/k^2} + f_2 \cdot \overline{m/k} + f_3 \cdot \overline{m}}{N[b_1 \cdot s + b_2 \cdot (1-s)]^2} \quad (6.25)$$

$$= \frac{b_1 \cdot (p_{f1} - p_{f2}) \cdot s}{N \cdot [b_1 \cdot s + b_2 \cdot (1-s)]} + \alpha \times \frac{f_1 \cdot \overline{m/k^2} + f_2 \cdot \overline{m/k} + f_3 \cdot \overline{m}}{N[b_1 \cdot s + b_2 \cdot (1-s)]^2} = 0 \quad (6.26)$$

$$\Rightarrow \quad \frac{b_2 \cdot (p_{f2} - p_{f1}) \cdot (1-s)}{N \cdot [b_1 \cdot s + b_2 \cdot (1-s)]} = \frac{b_1 \cdot (p_{f1} - p_{f2}) \cdot s}{N \cdot [b_1 \cdot s + b_2 \cdot (1-s)]} \quad (6.27)$$

$$\Rightarrow \quad \frac{(p_{f1} - p_{f2}) \cdot [b_1 \cdot s + b_2 \cdot (1-s)]}{N \cdot [b_1 \cdot s + b_2 \cdot (1-s)]} = 0 \quad (6.28)$$

$$\Rightarrow \quad (p_{f1} - p_{f2}) = 0. \quad (6.29)$$

Thus, we can have the following necessary condition for the ESS $[p^*_{f1}, p^*_{f2}]^T$, that is,

Necessary Condition of the ESS: If $[p_{f1}, p_{f2}]^T$ is ESS, then $p_{f1} = p_{f2}$ must hold. Plugging this necessary condition into either (6.23) or (6.24), we can have

$$
\begin{aligned}
&(1-p_{f1})\cdot p_{f1}\times \\
&[\Phi(\overline{m}-3\overline{m/k}+2\overline{m/k^2})p_{f1}+\Phi(\overline{m/k}-\overline{m/k^2})+\Phi_n(\overline{m}-\overline{m/k})] &= 0, \quad \text{or} \\
&(1-p_{f2})\cdot p_{f2}\times \\
&[\Phi(\overline{m}-3\overline{m/k}+2\overline{m/k^2})p_{f1}+\Phi(\overline{m/k}-\overline{m/k^2})+\Phi_n(\overline{m}-\overline{m/k})] &= 0.
\end{aligned} \tag{6.30}
$$

The above equations also shows three candidates for the ESS which are:

$$
\mathrm{ESS}_1 = \begin{cases} p^*_{f1} = 0 \\ p^*_{f2} = 0, \end{cases} \tag{6.31}
$$

$$
\mathrm{ESS}_2 = \begin{cases} p^*_{f1} = 1 \\ p^*_{f2} = 1, \end{cases} \tag{6.32}
$$

$$
\mathrm{ESS}_3 = \begin{cases} p^*_{f1} = -\dfrac{\Phi(\overline{m/k}-\overline{m/k^2})+\Phi_n(\overline{m}-\overline{m/k})}{\Phi[\overline{m}-3\overline{m/k}+2\overline{m/k^2}]} \\ p^*_{f2} = -\dfrac{\Phi(\overline{m/k}-\overline{m/k^2})+\Phi_n(\overline{m}-\overline{m/k})}{\Phi[\overline{m}-3\overline{m/k}+2\overline{m/k^2}]}. \end{cases} \tag{6.33}
$$

Among three ESSs, there are two trivial results where both p_{f1} and p_{f2} equal to 0 or 1. This two scenarios show that the whole population will finally not forward or forward the information, which are two extreme situations. As for ESS_3, it can be written as

$$
\begin{aligned}
p^*_{f1} = p^*_{f2} &= -\frac{\Phi\cdot(\overline{m/k}-\overline{m/k^2})+\Phi_n\cdot(\overline{m}-\overline{m/k})}{\Phi\cdot[\overline{m}-3\overline{m/k}+2\overline{m/k^2}]} \\
&= -\frac{\Phi\cdot(\overline{m/k}-\overline{m/k^2})+\Phi_n\cdot(\overline{m}-\overline{m/k})}{\Phi\cdot[(\overline{m}-\overline{m/k})-2(\overline{m/k}-\overline{m/k^2})]} \\
&= -\frac{\Phi\cdot\frac{\overline{m/k}-\overline{m/k^2}}{\overline{m}-\overline{m/k}}+\Phi_n}{\Phi\cdot[1-2\frac{\overline{m/k}-\overline{m/k^2}}{\overline{m}-\overline{m/k}}]}.
\end{aligned} \tag{6.34}
$$

With the same assumption as in [10] that k is a rather large value, we can have

$$
\overline{m}-\overline{m/k} = \sum_k \lambda(k)m(k)\left(1-\frac{1}{k}\right) \approx \sum_k \lambda(k)m(k) = \overline{m}, \quad \text{and} \tag{6.35}
$$

$$
\overline{m/k}-\overline{m/k^2} = \sum_k \lambda(k)\frac{m(k)}{k}\left(1-\frac{1}{k}\right) \approx \sum_k \lambda(k)\frac{m(k)}{k} = \overline{m/k} < \frac{\overline{m}}{k_{min}}. \tag{6.36}
$$

where k_{min} is the smallest degree, and is also assumed to be rather large. Thus, we have the approximation of ESS_3, that is,

$$p_{f1}^* = p_{f2}^* = -\frac{\Phi \frac{\overline{m/k} - \overline{m/k^2}}{\overline{m} - \overline{m/k}} + \Phi_n}{\Phi[1 - 2\frac{\overline{m/k} - \overline{m/k^2}}{\overline{m} - \overline{m/k}}]} \approx -\frac{\Phi_n}{\Phi} \tag{6.37}$$

Since p_{f1} and p_{f2} should both lie in the range [0, 1], ESS_3 is only feasible when $0 < -\frac{\Phi_n}{\Phi} < 1$, or equivalently $u_{fn} > u_{ff}$ and $u_{fn} > u_{nn}$.

Next, we examine the stability at each ESS candidates. The system dynamics in (6.21) and (6.22) are nonlinear. Thus, we adopt the stability analysis in [10–12]. That is, whether a ESS candidate is evolutionary stable can be obtained through analyzing the Jacobian matrix of the dynamics as follows:

$$J = \begin{pmatrix} \frac{\partial \dot{p}_{f1}}{\partial p_{f1}} & \frac{\partial \dot{p}_{f1}}{\partial p_{f2}} \\ \frac{\partial \dot{p}_{f2}}{\partial p_{f1}} & \frac{\partial \dot{p}_{f2}}{\partial p_{f2}} \end{pmatrix}. \tag{6.38}$$

In order to simplify the calculation and annotation, we denote

$$g_1 = b_2 \cdot (1 - s) \cdot (p_{f2} - p_{f1}), \tag{6.39}$$

$$g_2 = b_1 \cdot s \cdot (p_{f1} - p_{f2}), \tag{6.40}$$

$$h = b_1 \cdot s + b_2 \cdot (1 - s), \text{ and} \tag{6.41}$$

$$f = f_1 \cdot \overline{m/k^2} + f_2 \cdot \overline{m/k} + f_3 \cdot \overline{m}. \tag{6.42}$$

Only g_1, g_2 and f are the function of p_{f1} and p_{f2}. Thus, the system dynamics can be simplified to

$$\dot{p}_{f1} = \frac{g_1}{N \cdot h} + \alpha \times \frac{f}{N \cdot h^2}, \tag{6.43}$$

$$\dot{p}_{f2} = \frac{g_2}{N \cdot h} + \alpha \times \frac{f}{N \cdot h^2}. \tag{6.44}$$

Then, the Jacobian matrix can be simplified as

$$\begin{aligned} J &= \begin{pmatrix} \frac{\partial \dot{p}_{f1}}{\partial p_{f1}} & \frac{\partial \dot{p}_{f1}}{\partial p_{f2}} \\ \frac{\partial \dot{p}_{f2}}{\partial p_{f1}} & \frac{\partial \dot{p}_{f2}}{\partial p_{f2}} \end{pmatrix} = \frac{1}{N \cdot h} \begin{pmatrix} J_1 & J_2 \\ J_3 & J_4 \end{pmatrix} \\ &= \frac{1}{N \cdot h} \begin{pmatrix} \frac{\partial g_1}{\partial p_{f1}} + \frac{\alpha}{h} \cdot \frac{\partial f}{\partial p_{f1}} & \frac{\partial g_1}{\partial p_{f2}} + \frac{\alpha}{h} \cdot \frac{\partial f}{\partial p_{f2}} \\ \frac{\partial g_2}{\partial p_{f1}} + \frac{\alpha}{h} \cdot \frac{\partial f}{\partial p_{f1}} & \frac{\partial g_2}{\partial p_{f2}} + \frac{\alpha}{h} \cdot \frac{\partial f}{\partial p_{f2}}. \end{pmatrix} \end{aligned} \tag{6.45}$$

We first give the partial derivatives of g_1 and g_2 with respect to p_{f1} and p_{f2}

$$\frac{\partial g_1}{\partial p_{f1}} = -b_2(1-s), \quad \frac{\partial g_1}{\partial p_{f2}} = b_2(1-s),$$
$$\frac{\partial g_2}{\partial p_{f1}} = b_1 s, \quad \text{and} \quad \frac{\partial g_2}{\partial p_{f2}} = -b_1 s. \tag{6.46}$$

The evolutionary stability requires that $J_1 + J_4 < 0$ and $det(J) > 0$. J_1 is the summation of two terms, which are $-b_2(1-s)$ and $\frac{\alpha}{h} \cdot \frac{\partial f}{\partial p_{f1}}$, respectively. The second term is multiplied by $\frac{\alpha}{h}$ which is a rather small value in the weak selection scenario, and thus can be omitted. The first term $-b_2(1-s)$ is guaranteed to be negative. Therefore J_1 is negative. J_2 can be proved to be negative in a similar way. Thus, the first condition is satisfied.

The second condition can be calculated as

$$\begin{aligned} det(J) =& \left(\frac{\partial g_1}{\partial p_{f1}} + \frac{\alpha}{h} \cdot \frac{\partial f}{\partial p_{f1}}\right) \cdot \left(\frac{\partial g_2}{\partial p_{f2}} + \frac{\alpha}{h} \cdot \frac{\partial f}{\partial p_{f2}}\right) \\ &- \left(\frac{\partial g_1}{\partial p_{f2}} + \frac{\alpha}{h} \cdot \frac{\partial f}{\partial p_{f2}}\right) \cdot \left(\frac{\partial g_2}{\partial p_{f1}} + \frac{\alpha}{h} \cdot \frac{\partial f}{\partial p_{f1}}\right) \\ =& \frac{\partial g_1}{\partial p_{f1}} \cdot \frac{\partial g_2}{\partial p_{f2}} \\ &- \frac{\partial g_1}{\partial p_{f2}} \cdot \frac{\partial g_2}{\partial p_{f1}} + \frac{\alpha}{h}\left[\frac{\partial f}{\partial p_{f1}}\left(\frac{\partial g_2}{\partial p_{f2}} - \frac{\partial g_1}{\partial p_{f2}}\right) + \frac{\partial f}{\partial p_{f2}}\left(\frac{\partial g_1}{\partial p_{f1}} - \frac{\partial g_2}{\partial p_{f1}}\right)\right] \\ =& -\alpha\left(\frac{\partial f}{\partial p_{f1}} + \frac{\partial f}{\partial p_{f2}}\right) \end{aligned} \tag{6.47}$$

Considering the necessary condition for the ESS, that $p_{f1} = p_{f2}$, We can calculate $det(J)$ at $p_{f1} = p_{f2}$. Therefore, we can obtain

$$\begin{aligned} det(J) =& (b2 \cdot (s-1) - b1 \cdot s) \cdot \\ & \{\Phi \cdot [(1 - 6p_{f1} + 6p_{f1}^2)(\overline{m/k^2} - \overline{m/k}) \\ &+ (2p_{f1} - 3p_{f1}^2)(\overline{m/k} - \overline{m})] + \Phi_n \cdot (1 - 2p_{f1})(\overline{m/k} - \overline{m})\}. \end{aligned} \tag{6.48}$$

Under the assumption that k is rather large, plugging $p_{f1} = 0$, 1 and $-\frac{\Phi_n}{\Phi}$ into the above expression yields

$$\begin{aligned} det(J)|_{p_{f1}=0} =& (b2 \cdot (s-1) - b1 \cdot s) \cdot [(\overline{m/k^2} - \overline{m/k}) \cdot \Phi + (\overline{m/k} - \overline{m}) \cdot \Phi_n] \\ \approx& \overline{m} \cdot (b2 \cdot (1-s) + b1 \cdot s) \cdot \Phi_n \\ =& \overline{m} \cdot (b2 \cdot (1-s) + b1 \cdot s) \cdot (u_{fn} - u_{nn}), \\ det(J)|_{p_{f1}=1} =& (b2 \cdot (s-1) - b1 \cdot s) \times \end{aligned}$$

$$
\begin{aligned}
&\{[(\overline{m}-\overline{m/k})-(\overline{m/k}-\overline{m/k^2})]\cdot\Phi+(\overline{m}-\overline{m/k})\cdot\Phi_n\}\\
&\approx -\overline{m}\cdot(b2\cdot(1-s)+b1\cdot s)\cdot(\Phi_n+\Phi)\\
&= -\overline{m}\cdot(b2\cdot(1-s)+b1\cdot s)\cdot(u_{ff}-u_{fn}),
\end{aligned}
\tag{6.49}
$$

$$
det(J)|_{P_{f1}=\frac{-\Phi_n}{\Phi}} \approx \overline{m}\cdot(b2\cdot(1-s)+b1\cdot s)\cdot\left(\frac{-\Phi_n}{\Phi}\right)\cdot(\Phi+\Phi_n).
$$

According to the value of $det(J)|_{P_{f1}=0}$, $det(J)|_{P_{f1}=1}$ and $det(J)|_{P_{f1}=\frac{-\Phi_n}{\Phi}}$, the stability of three ESSs can be discussed in the following cases:

- $u_{ff} > \max(u_{fn}, u_{nn})$. In this case, $det(J)|_{P_{f1}=1} < 0$. Thus, ESS_2 is stable.
- $u_{nn} > \max(u_{ff}, u_{fn}$. In this case $det(J)|_{P_{f1}=0} < 0$. Thus, ESS_1 is stable.
- $u_{fn} > \max(u_{ff}, u_{nn})$. In this case, both $det(J)|_{P_{f1}=0} > 0$ and $det(J)|_{P_{f1}=1} > 0$. Thus, both ESS_1 and ESS_2 are unstable. However, in this case, ESS_3 lies in the range [0, 1] and $det(J)|_{P_{f1}=\frac{-\Phi_n}{\Phi}} < 0$, which indicates that ESS_3 is stable.

In summary, we can obtain the overall ESS in (6.13).

References

1. J. Yang and J. Leskovec, "Patterns of temporal variation in online media," in *Proceedings of the 4th ACM International Conference on Web Search and Data Mining*, 2011, pp. 177–186
2. J. Yang and J. Leskovec, "Modeling information diffusion in implicit networks," in *Proceedings of IEEE International Conference on Data Mining*, 2010, pp. 599–608
3. S. A. Myers and J. Leskovec, "Clash of the contagions: Cooperation and competition in information diffusion," in *Proceedings of IEEE International Conference on Data Mining*, 2012, pp. 539–548
4. C. Yang, J. Tang, M. Sun, G. Cui, and Z. Liu, "Multi-scale information diffusion prediction with reinforced recurrent networks," in *Proceedings of International Joint Conference on Artificial Intelligence*, 2019, pp. 4033–4039
5. D.J. Daley, D.G. Kendall, Epidemics and rumours. Nature **204**(4963), 1118 (1964)
6. K. Lerman and R. Ghosh, "Information contagion: An empirical study of the spread of news on digg and twitter social networks," in *International Conference on Weblogs and Social Media*, 2010
7. Q. Xu, Z. Su, K. Zhang, P. Ren, X. Shen, Epidemic information dissemination in mobile social networks with opportunistic links. IEEE Transactions on Emerging Topics in Computing **3**(3), 399–409 (2017)
8. M. Granovetter, Threshold models of collective behavior. American Journal of Sociology **83**(6), 1420–1443 (1978)
9. J. Goldenberg, B. Libai, E. Muller, Talk of the network: A complex systems look at the underlying process of word-of-mouth. Marketing Letters **12**(3), 211–223 (2001)
10. C. Jiang, Y. Chen, K.J.R. Liu, Graphical evolutionary game for information diffusion over social networks. IEEE Journal of Selected Topics in Signal Processing **8**(4), 524–536 (2014)
11. C. Jiang, Y. Chen, K.J.R. Liu, Evolutionary dynamics of information diffusion over social networks. IEEE Transactions on Signal Processing **62**(17), 4573–4586 (2014)
12. X. Cao, Y. Chen, C. Jiang, K.J.R. Liu, Evolutionary information diffusion over heterogeneous social networks. IEEE Transactions on Signal and Information Processing over Networks **2**(4), 595–610 (2016)

13. Y. Li, B. Qiu, Y. Chen, and H. V. Zhao, "Analysis of information diffusion with irrational users: A graphical evolutionary game approach," in *Proceedings of IEEE International Conference on Acoustics, Speech and Signal Processing*, 2019, pp. 2527–2531
14. K. Leetaru, "Twitter users mostly retweet politicians and celebrities. Thats a big change," *the Washington Post*, 2019, mar. 8
15. Y. Liu, B. Wang, B. Wu, S. Shang, Y. Zhang, C. Shi, Characterizing super-spreading in Microblog: An epidemic-based information propagation model. Physica A: Statistical Mechanics and its Applications **463**, 202–218 (2016)
16. D. Kempe, J. Kleinberg, and É. Tardos, "Maximizing the spread of influence through a social network," in *Proceedings of ACM SIGKDD International Conference on Knowledge Discovery and Data Mining*, 2003, pp. 137–146
17. A. Zareie, A. Sheikhahmadi, M. Jalili, Influential node ranking in social networks based on neighborhood diversity. Future Generation Computer Systems **94**, 120–129 (2019)
18. D.J. Watts, P.S. Dodds, Influentials, networks, and public opinion formation. Journal of Consumer Research **34**(4), 441–458 (2007)
19. R. Pastor-Satorras, A. Vázquez, A. Vespignani, Dynamical and correlation properties of the internet. Physical review letters **87**(25), 258701 (2001)
20. M. Catanzaro, M. Boguná, R. Pastor-Satorras, Generation of uncorrelated random scale-free networks. Physical review e **71**(2), 027103 (2005)
21. D. Yao, P. van der Hoorn, and N. Litvak, "Average nearest neighbor degrees in scale-free networks," arXiv preprint arXiv:1704.05707, 2017

Chapter 7
Concluding Remarks

Abstract The crowd networks include millions of deeply connected individuals, smart devices, government agencies, organizations, and enterprises, who actively interact with each other and influence each other's decisions. This book provides a holistic framework to study their decision-making processes and their interactions and analyze its impact on the crowd networks. This investigation offers critical guidelines on crowd networks' design to avoid detrimental events that affect our society and economy. This chapter reviews essential findings in each chapter and points out a few possible future directions. We aim to encourage researchers from different disciplines to address the challenging issues and explore the untouched territories in this emerging research field.

Keywords Crowd networks · Evolutionary game theory · Information diffusion · Evolutionary dynamics

In the emerging crowd cyber-eco systems, millions of deeply connected individuals, smart devices, government agencies, and enterprises actively interact with each other and influence each other's decisions. It is crucial to understand such intelligent entities' behaviors, and to study their strategic interactions, and to understand its impact on the complex networks. This investigation provides important guidelines on the design of reliable systems capable of predicting and preventing detrimental events with adverse effects on our society and economy.

This book offers a holistic framework to study behavior and evolutionary dynamics in large-scale, decentralized, and heterogeneous crowd networks. It reviews the fundamental methodologies to analyze user interactions and evolutionary dynamics in crowd networks, and covers recent advances in this emerging inter-disciplinary research direction. Using information diffusion over social networks as an example, it provides a thorough investigation of the impact of user behavior on the network evolution process. It demonstrates how such an understanding can help improve network performance.

Evolutionary game theory offers important tools to model and understand user dynamics in complex, heterogeneous and decentralized networks, where players (users) in the network constantly interact with each other and influence each other's decisions. Evolutionary game theory has been used to understand behavior dynamics

Y. Chen and H. V. Zhao, *Behavior and Evolutionary Dynamics in Crowd Networks*, Lecture Notes in Social Networks, https://doi.org/10.1007/978-981-15-7160-2_7

in many applications, such as natural, biological, and social networks. In this book, we use evolutionary game theory to model the information diffusion process in social networks, and review in Chap. 2 fundamentals of graphical evolutionary game theory and its application in information diffusion.

Though social networks fundamentally change our social life and become a major media for us to interact with others and share experience and knowledge. However, it also facilitates the fast spread and possible outbreak of misleading, false, and harmful information. Thus, it is important to study such "irrational" and malicious behavior in social networks, and understand how they influence other rational users' decisions. Chapter 3 uses graphical evolutionary game theory to study the impact of such malicious behavior, and analyzes the corresponding population dynamics and the evolutionary stable states.

To address the negative impact of these "irrational" and malicious behavior on others as well as the entire network, it is crucial to design countermeasures that detect and prevent such malicious behavior. Chapter 4 introduces a reputation-based mechanism and proposes a smart evolution model based on indirect reciprocity. Given the proposed reputation mechanism, including social norm and reputation updating policy, we theoretically analyze the evolutionary dynamics and the corresponding evolutionary stable states. Simulation results on synthetic and Facebook networks show that the proposed reputation mechanism can effectively mitigate the negative impact of malicious users.

Note that most prior works consider the propagation of one single message over social networks, while in reality, correlated information often spread together and influence each other. Chapter 5 addresses this challenge, and extends the previous graphical evolutionary game-theoretic framework to study the diffusion of multi-source correlated information over social networks. We theoretically analyze the population dynamics, relationship dynamics, and influence dynamics there, validate the performance of the proposed model on synthetic and real Facebook networks, and test on real data crawled from Sina Weibo.

In social networks, there exist some "super users" whose social standings are higher than others, and they often have a significant influence on others. Chapter 6 extends the graphical evolutionary game-theoretic framework to study the impact of such "super users" on the information diffusion process. Interestingly, our investigation shows that super users themselves can not change the evolutionarily stable state of information diffusion, but only the evolutionary dynamics. At the same time, strategy update rules are critical to the evolutionary stable states of the diffusion process.

Crowd network is an emerging and inter-disciplinary research field, and there are many untouched territories that are still waiting for us to explore. For example, the propagation of information not only depends on how users interact with each other, but also the content of the message. Thus, it is critical to jointly analyze the content of each message as well as analyze the interplay among users to better understand the dynamic spreading of information over social networks. We aim to encourage researchers from different disciplines, including but limited to computer science, networking, big data, complex systems, and economics, to collectively address many challenging issues and ultimately to design networks with efficient, effective, and reliable services.

Index

Y. Chen and H. V. Zhao, *Behavior and Evolutionary Dynamics in Crowd Networks*, Lecture Notes in Social Networks, https://doi.org/10.1007/978-981-15-7160-2

G

H

I

K

L

M

N

O

P

Printed in the United States
by Baker & Taylor Publisher Services